The Prentice-Hall Science Education Series

This series reflects an ongoing commitment to science education, offering teachers, students, and the general public critical information, instruction, and activity resources that activate effective and enjoyable learning in science and mathematics. The series is designed for use by children of all ages and the adults who work with them in both school and non-school settings. These practical science handbooks give readers basic concepts, ideas, and applications to meet the challenges of the future. Among the titles:

Teaching Children About Science:
Ideas and Activities Every Teacher and Parent Can Use
Elaine Levenson

Teaching the Fun of Physics: 101 Activities
to Make Science Education Easy and Enjoyable
Janice Pratt VanCleave

The Earth and How It Works: A Lab Manual and Workbook with
Teaching Ideas, Projects, and Activities in Environmental Science
Philip R. Holzinger

The Rocket Book: A Guide to Building and Launching
Model Rockets for the Space Age
Robert L. Cannon and Michael A. Banks

ROBERT L. CANNON
AND
MICHAEL A. BANKS

THE ROCKET BOOK

A Guide to Building and Launching Model Rockets for the Space Age

Prentice-Hall, Inc., Englewood Cliffs, N.J. 07632

Library of Congress Cataloging in Publication Data

Cannon, Robert L.
The rocket book.

(The Prentice-Hall science education series)

Bibliography: p.
Includes index.
1. Rockets (Aeronautics)—Models. I. Banks, Michael A. II. Title. III. Series.
TL844.C37 1985 629.47'5 85–587
ISBN 0–13–782244–8

10 9 8 7 6 5 4 3 2
Printed in the United States of America

Editorial/production supervision: Joe O'Donnell Jr.
Book design: Maria Carella
Cover design: Hal Siegel
Manufacturing buyer: Carol Bystrom

ISBN 0-13-782244-8

Front cover photo courtesy of Estes Industries/Hi-Flier Mfg. Co.
Most of the photographs and drawings in this book were provided courtesy of Estes Industries.

★ *This book is dedicated to Vern Estes, who recognized the educational potential of model rocketry, and to Bill Simon, who helped this potential be realized.*

CONTENTS

FOREWORD

Since the development of the model rocket engine in 1957 and the mass marketing of model rocket kits and engines beginning in 1960, the model rocket has been available as a versatile classroom tool. I say "versatile" because the model rocket can be used in the mathematics classroom, the science classroom, in industrial arts, and even in arts and crafts classrooms.

To fully understand the motion of a model rocket requires a foundation in calculus. To understand methods of measuring a rocket's altitude requires understanding of basic geometry and trigonometry. To understand the interplay of forces and movements governing model rocket motion requires knowledge of physics, and to understand the internal ballistics of the model rocket engine requires knowledge of chemistry.

Building a model rocket launch system or an altitude tracking device is a project for an industrial arts classroom. Constructing a model rocket kit with paints and decals is an arts and crafts project. The model rocket can fit into almost any lesson plan in any classroom. Indeed, it is a versatile educational tool.

Model rockets even show up in extracurricular activities within afterschool science and computer clubs. Many schools have model rocket clubs as extracurricular activities. These clubs organize regular contests, demonstrations before the school student body and audiences in other schools, meetings with aerospace films and lectures, plus a multitude of hobby activities for the participating students.

You, as a teacher, take a risk bringing model rocketry into your classroom. You risk the possibility of exciting your students to want to learn more about mathematics and science and to want to learn more about constructing scientific apparatus to launch model rockets and to measure their performance. You risk the possibility that your students will want to form an afterschool model rocket club, complete with fun flys and spirited contests.

As a teacher, you have an opportunity to learn more about model rockets and model rocketry. This opportunity is available to you within the pages of this book. You can learn how to bring the model rocket into your classroom as a versatile tool. It is most certainly worth your while to do so, and your students will be the ones who benefit most.

J. PATRICK MILLER, president,
National Association of Rocketry

PREFACE

Welcome to the world's greatest hobby and educational activity—model rocketry!

Whether model rocketry is entirely new to you or you are an experienced model rocketeer, we have written this book for you. In this book, you will find all the information you need to get started in model rocketry activities, along with valuable ideas and suggestions for using model rocketry in the classroom.

The applications of model rocketry are not limited to the science classroom. Model rocketry can be, with planning, a useful supplement to studies in mathematics, physics, industrial arts, language arts and communication, and more.

Model rocketry can also be a useful learning incentive and enrichment activity. It has been used successfully in programs for the gifted, the learning disabled and handicapped, and mainstream students at virtually all grade levels, from elementary school through college.

In writing this book, we have drawn upon our extensive knowledge and experience as educators and enthusiastic model rocketeers. Our goal has been not only to inform you of the educational value of model rocketry, but also, we hope, to pass along some of our enthusiasm for the hobby. We hope that you will not only find enthusiasm in the learning experiences that model rocketry offers your students, but also that you will find this same enthusiasm for yourself. Indeed, you may find, as so many thousands have before you, that "model rocketry is habit-forming!"

There is no way to acknowledge the contributions of the many individuals who have helped model rocketry become such an effective aid to learning for well over one million youngsters. We would, however, like to acknowledge a few individuals who have been responsible for initiating the development of educational programs in model rocketry and those who have encouraged and contributed to this development:

G. Harry Stine and Orville Carlisle, who made the hobby possible; Vern and Gleda Estes, who made model rocketry available to everyone and who made making happy, successful model rocketeers the basis of the business they founded; Bill Simon, who hired one of the authors and encouraged and helped him in developing and improving the educational program already started by Estes Industries; and the hundreds of teachers and youth workers who have shared their creative ideas on how to use model rocketry through their articles published on *Estes's Educator News*.

1 INTRODUCTION TO MODEL ROCKETRY

Model rocketry is many things at once—a sport, a hobby, a science, a research tool, and an education aid. Millions of model rocketeers of all ages have flown model rockets since the inception of the hobby in 1957, and many who were building and flying these miniature rockets then are still flying them today.

Model rocketry is a great hobby. It permits you to design, build, and safely launch rocket-powered vehicles.

Model rockets are built using lightweight materials, such as paper, wood, and plastic. No structural parts may be made of metal. The rocket engines that power model rockets are preloaded, factory-made engines. The rocketeer does not make propellants. The entire model rocket, including the engine, is very light. The model rocket engine contains only a very small amount of propellant.

Model rockets are brought safely back to earth by recovery systems, such as parachutes, to be flown again and again.

THREE TYPES OF ROCKETEERS

★ The three types of rocketeers are professional, amateur, and model.

The most famous type of rocketeer is the *professional*. These are the men and women who make their living through the use of rockets. The astronauts (see Figure 1–1) are the most famous professional rocketeers.

Figure 1-1. Astronauts. Photo courtesy NASA.

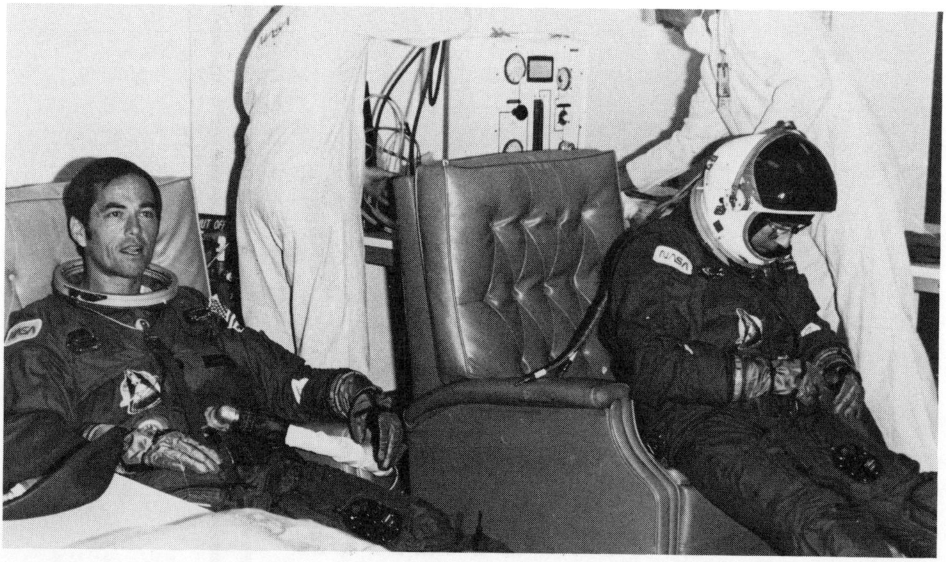

Figure 1-2. Amateur rocketeers.

Most professional rocketeers are not famous. The thousands of men and women who design the rocket engines, the rocket vehicles, the guidance systems, the radios, the telemetry devices, the launch pads, and the numerous systems and items necessary for rocket flight seldom make the TV news. However, their contributions are essential to the success of each rocket flight. They do the dreaming, the planning, and the testing to create new rockets. They work on the spaceships of today and tomorrow.

Many thousands of people work in the machine shops, the electrical shops, the propellant factories, the assembly lines, and dozens of other places. Their efforts all contribute to a successful flight. Everyone who works in the space program must perform quality work. As one of the Apollo 11 astronauts humorously expressed it, "It gives you a great feeling of confidence to realize that you are sitting on top of millions of separate parts, each of which was built by the lowest bidder." Every individual part must work in the right way at the right time, or the flight may end in disaster. This is why so many checks and doublechecks are made on each item that goes into a rocket.

A second type of rocketeer is the *amateur rocketeer*. (See Figure 1–2.) These individuals make their own rocket propellants. They also make their own rocket bodies, often from metal. Some of these amateur rocketeers believe that they practice a safe form of rocketry. The American Institute of Aeronautics and Astronautics does not agree. A study they conducted indicated that rocketeers who made their own propellants had one chance in seven of killing themselves or causing an accident so serious that they would be maimed for life for each year that they practiced amateur rocketry.

There have been a few successful amateurs who worked under the supervision of qualified experts. Unfortunately for those who would like to participate in amateur rocketry experiments, there are few qualified experts. The accessibility of the few qualified experts has always been extremely limited. In addition to these limitations, the availability of suitable areas to work in has been very restricted. Few programs of properly supervised amateur rocketry have been conducted. We are not aware of any suitable current programs of this nature.

The third type of rocketeer is the *model rocketeer*. These individuals practice their hobby in safety, because they follow the Model Rocketry Safety Code (see page 6).

Model rocketeers can launch small rockets for fun; can send high-performance rockets thousands of feet into the air; can launch rockets that stage themselves automatically in mid-flight, just like the huge NASA rockets; can take excellent photographs in color from high in the air; and can launch a wide variety of payloads, including functioning telemetry systems (see Figures 1-3A and 1-3B).

Figure 1-3A. Putting a rocket on launch pad.

Figure 1-3B. Launching a model rocket.

MODEL ROCKETRY SAFETY CODE

1. Construction. My model rockets will be made of lightweight materials such as paper, wood, plastic, and rubber, without any metal as structural parts.

2. Engines. I will use only preloaded, factory-made N. A. R. Safety Certified model rocket engines in the manner recommended by the manufacturer. I will not change in any way nor attempt to reload these engines.

3. Recovery. I will always use a recovery system in my rockets that will return them safely to the ground so that they may be flown again.

4. Weight Limits. My model rocket will weigh no more than 453 grams (16 ounces) at liftoff, and the engine will contain no more than 113 grams (4 ounces) of propellant.

5. Stability. I will check the stability of my model rockets before their first flight, except when launching models of already proven stability.

6. Launching Systems. The system I use to launch my model rockets must be remotely controlled and electrically operated, and will contain a switch that will return to "off" when released. I will remain at least 15 feet from any rocket that is being launched.

7. Launch Safety. I will not let anyone approach a model rocket on a launcher until I have made sure that either the safety interlock key has been removed or the battery has been disconnected from my launcher.

8. Flying Conditions. I will not launch my model rockets in high winds, near buildings, power lines, tall trees, low-flying aircraft, or under any conditions which might be dangerous to people or property.

9. Launch Area. My model rockets will always be launched from a cleared area, free of any easy-to-burn materials, and I will only use nonflammable recovery wadding in my rockets.

10. Jet Deflector. My launcher will have a jet deflector device to prevent the engine exhaust from hitting the ground directly.

11. Launch Rod. To prevent accidental eye injury, I will always place the launcher so the end of the rod is above eye level or cap the end of the rod with my hand when approaching it. I will never place my head or body over the launching rod. When my launcher is not in use, I will always store it so that the launch rod is not in an upright position.

12. Power Lines. I will never attempt to recover my rocket from power lines or other dangerous places.

13. Launch Targets and Angle. I will not launch rockets so their flight path will carry them against targets on the ground, and will never use an explosive warhead nor a payload that is intended to be flammable. My launching device will always be pointed within 30 degrees of vertical.

14. PreLaunch Test. When conducting research activities with unproven designs or methods, I will, when possible, determine their reliability through prelaunch tests. I will conduct launchings of unproven designs in complete isolation from persons not participating in the actual launching.

This Model Rocketry Safety Code is approved by the National Association of Rocketry (N. A. R.) and the Hobby Industry Association of America (H. I. A. A.).

The activities of model rocketeers are very similar to those of professional rocketeers in many respects. Both professional and model rocketeers build and fly established designs and experimental rockets, and both follow prescribed safety routines. Many types of professional rockets use solid fuel. (The space shuttle's boosters use solid fuel.) Model rockets are subject to and make use of the same aerodynamic and physical principles as "full-size" rockets; the major differences are size and cost.

Formulated by experienced rocketeers and evolved with model rocketry, the Safety Code should be followed in each and every model rocketry activity. Memorize it and use it.

GETTING STARTED IN MODEL ROCKETRY

Most model rocketeers start by purchasing a model rocket kit. (See Figure 1–4A.) Putting it together is a challenging, fun project.

Then comes the launch!

The rocketeer takes the completed rocket, a package of engines, a

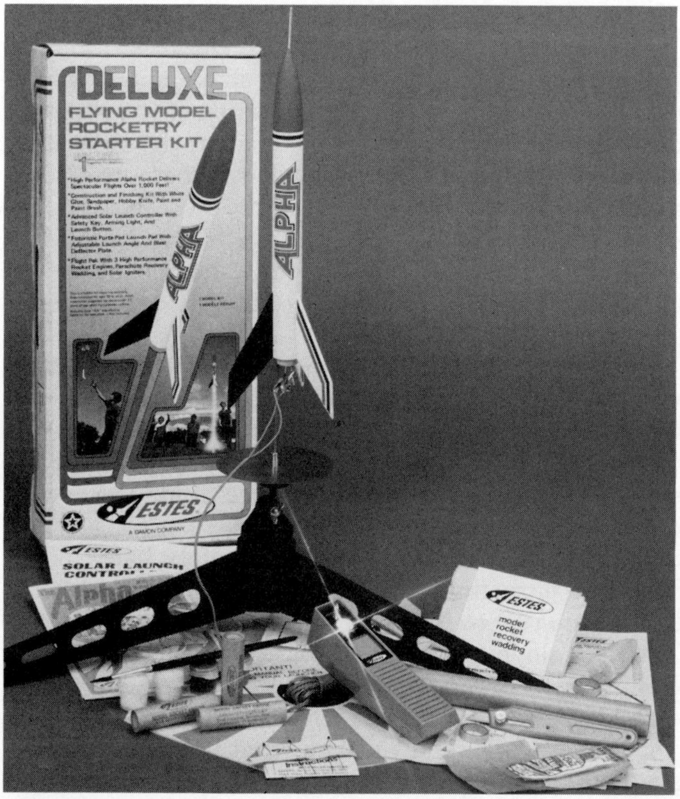

Figure 1-4A. Starter set.

package of recovery wadding, the launch pad, and the electrical ignition system, and heads for the nearest large, open area. This sounds like a lot of gear to carry, but it really isn't. Everything, except possibly the rocket, is miniaturized. If the site chosen is on private property, the rocketeer should be sure to get the owner's permission before launching.

The launch pad is set up. The rocket's recovery system is reprepped. The electrical igniter is carefully placed in the engine. The engine is secured in the rocket. The rocketeer checks the area to make certain that all safety rules are met. Then the ignition key is inserted. The continuity light comes on. Now the countdown!

5...4...3...2...1...Blast-off!

The rocket streaks into the sky. A thin trail of smoke appears as the expectancy mounts. Then the 'chute! Mission accomplished!

A TYPICAL MODEL ROCKET

★ Model rockets come in a wide variety of sizes, shapes, and types. No matter what the shape, however, they all share certain characteristics, as shown in Figure 1–4B.

The basic structure of the rocket that gives it its shape and to which other parts are attached is called the *body tube*. This portion of the rocket is usually long and slender. Most body tubes are made of paper, tightly wound in a spiral pattern. The tube is strong yet light. The body tube may also be called the fuselage, or airframe, of the rocket.

The *nose cone* is the front end of the rocket. It is usually shaped to provide minimum resistance, or drag, to the air that the rocket moves through. The shock cord, and often the parachute, are attached to the nose cone.

The *fins* provide stability and guidance for the rocket in its flight. Fins may be made of balsa, cardboard, thin plywood, or plastic.

The *launch lug* guides the rocket during the fraction of a second after engine ignition. The launch lug is a small tube shaped like a soda straw, usually made of paper, which fits over the launch rod to guide the rocket during the first few feet of its flight.

The *recovery system* brings the rocket gently and safely back to the ground. It usually consists of a parachute attached to a shock cord, which is anchored to the body tube of the rocket. The shock cord absorbs much of the force of the recovery system deployment when the ejection charge funtions. There are several different types of recovery systems. Most recovery systems are stored in the rocket's body during the thrust and coast phases of each flight. All cause the rocket to descend slowly so that it is not damaged when it reaches the ground.

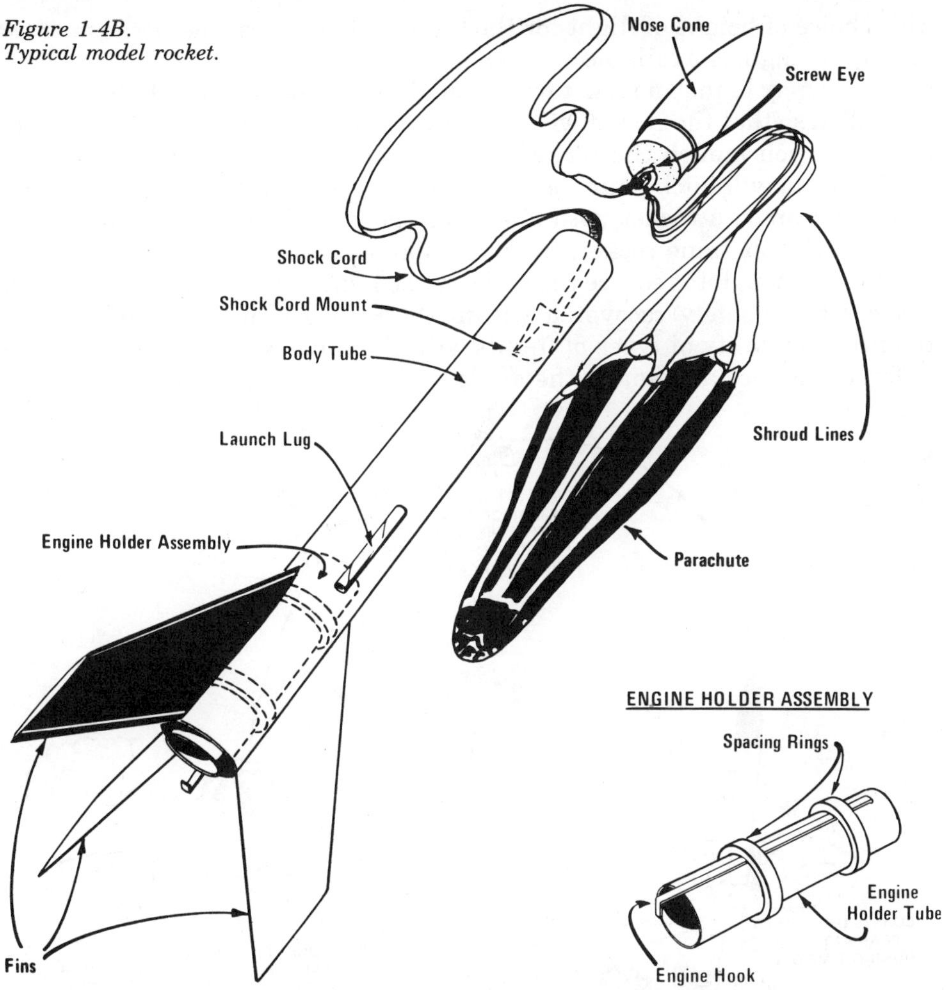

Figure 1-4B.
Typical model rocket.

The *engine* provides the power that causes the rocket to move. The *engine mount* holds the engine in the proper position in the body tube.

MODEL ROCKET FLIGHT PROFILE

★ Your model rocket is built to perform in a certain way. Each part must function properly if the model rocket is to fly safely and well.

The fins are not adjustable, so they must be the right size and shape and in the right position to guide the flight of the model rocket. You may launch the same rocket a hundred feet into the air or over a thousand feet.

This choice of height of flight for the same rocket is possible because of the variety of engines available.

You may launch a rocket from a small field. The rocket's flight must be controlled so that the rocket will return within the confines of the field. The next time you launch the rocket may be from the middle of a huge, open field. You may launch your rocket as high as you can this time. One time you may launch a rocket with absolutely no wind blowing. The next time you launch this same rocket, the wind may be blowing quite strongly, yet not strong enough to require the rocket to be launched at an angle of over 30 degrees to be safely recovered within the recovery area. To provide the required flexibility of types of flights with the same rocket, you will choose different engines and adjust the angle of the launch rod. (See Figure 1–5.)

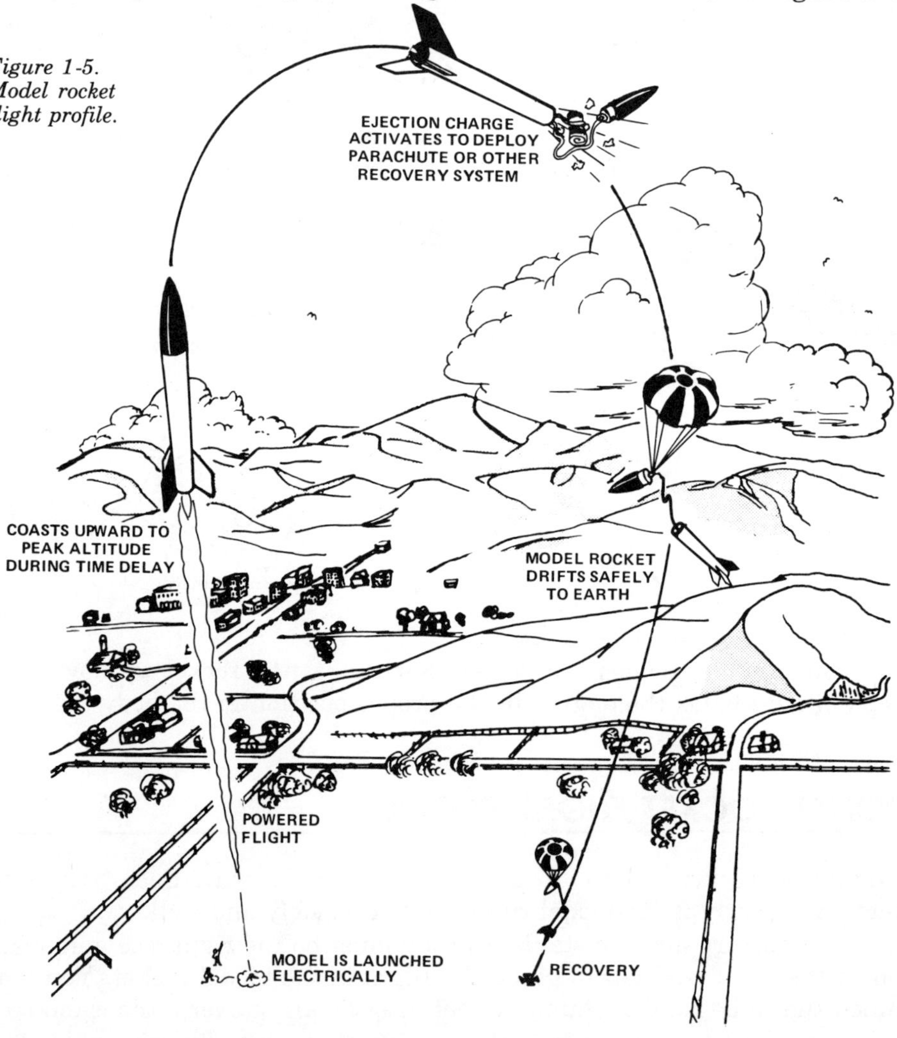

Figure 1-5. Model rocket flight profile.

The launch pad is very important to a successful model rocket flight. Prior to flight, the rocket is held securely in position as it sits on the launch pad. The first few feet of flight are controlled by the direction in which the launch rod is pointed.

THRUST PHASE

The flight begins when the model is launched by the ignition of the engine. The powered phase of the flight ends when the engine has consumed all its propellant.

During the powered portion of the flight, the rocket is undergoing thrust. It moves in response to the thrust, gravity, aerodynamic forces of air moving past the rocket and wind.

COASTING PHASE

As soon as the propellant is used up, the delay element is ignited for all engines except booster engines. The delay element burns for a set length of time to act as a timing device to control deployment of the recovery system. Recovery system deployment should occur at apogee, the highest point in the flight.

While the rocket is coasting, it is rapidly slowing down because of the forces of gravity and drag. The engine is not producing thrust during this phase of the flight, but the smoke-tracking and delay element of the engine is producing smoke. This smoke is extremely useful in tracking the rocket. Accurately observing, or even seeing, a small rocket a thousand or more feet above you and moving directly away from you is extremely difficult! Quite often, all you really see is the tracking smoke. This smoke trail permits you to follow the flight of the rocket.

RECOVERY PHASE

The engine's ejection charge deploys the recovery system when the smoke-tracking and delay element is used up. This should occur at apogee. The rocket drifts safely back to earth using the recovery system after the ejection charge activates it.

The recovery phase of the flight is the hardest to control. If you made an error in the construction or preparation of the recovery system, your rocket's time of descent may be approximately that of the ascent! This not only ruins your rocket, but it could be dangerous to spectators. Any model rocket flight that does not end in the successful recovery of the rocket is a failure.

Once the rocket's recovery system has been deployed by the activation of the ejection charge, the rocket is subject only to the force of gravity and the aerodynamic forces produced by the wind and its own surface areas.

THE MODEL ROCKET ENGINE

★ A model rocket engine is a sophisticated miniature solid-fuel device that has several functions.

Model rocket engines are factory-made devices that meet safety and performance standards established by the National Association of Rocketry, the Hobby Industry Association of America, and the National Fire Protection Association. (See Figure 1–6.) Model rocket engines also meet the standards established by the Food and Drug Administration and the Department of Transportation. They are safe to handle, store, and use. The design used in the construction of model rocket engines has been carefully worked out and modified over a period of over 25 years. Since 1958, when model rocketry was invented, over 200 million commercially manufactured model rocket engines have been safely and successfully used.

The casing of the typical model rocket engine is a cylinder of tightly rolled paper. At the base of the engine is the nozzle. The nozzle is made of

Figure 1-6.
Model rocket engine.

clay, a heat-resistant, rigid material. The propellant is above the nozzle. Above the propellant is the smoke-tracking and delay element, which is activated at propellant burnout and provides both a delay to permit the rocket to coast to apogee and a smoke trail to aid in tracking it. Above the delay element is the ejection charge, which produces hot gases that expand and blow away the clay cap at the top of the engine. The gases then expand into the body tube of the rocket and push the recovery system forward and out of the rocket.

Model rocket engines are manufactured in a range of sizes, but all have essentially the same shape. Model rocket engines also come in over 30 varieties, each one varying in total impulse, average impulse, thrusting time, and time of delay between propellant burnout and activation of the ejection charge.

Some companies print the information on model rocket engines in black. Estes Industries prints the information in color to supply additional information at a glance. Green print indicates that the engine is used for single-stage flights. Red print indicates that the engine is a booster engine. Booster engines have no delay/smoke-tracking elements or ejection charges. Purple or blue print indicates that the engine has an extremely long delay and should be used only for the upper stage of multistage rockets or for very light, high-performance model rockets.

The printing on the engine identifies the manufacturer of the engine and the type of engine. Additional information, such as instructions on disposal of the engine, DOT (Department of Transportation) classification, and the date of the engine's manufacture, often coded, is printed on the engine.

The engine code on each engine consists of a letter, a number, and another number preceded by a dash. The letter indicates the total impulse (total power) provided by that engine. The first number indicates the average thrust of the engine. This value is in Newtons. The last number is the delay time in seconds between the time of the engine's burnout and the activation of the ejection charge. (See Figure 1–7.)

Figure 1-7.
Total Impulse Classification

Code	*Pound-Seconds*	*Newton-Seconds*
¼A	0.00–0.14	0.00–0.625
½A	0.14–0.28	0.625–1.25
A	0.28–0.56	1.25–2.50
B	0.56–1.12	2.50–5.00
C	1.12–2.24	5.00–10.00
D	2.24–5.00	10.00–20.00

A Newton is the measurement of force required to move one kilogram of mass at a rate of one meter per second. The average thrust of an engine in Newtons multiplied by the thrust duration equals the total thrust in Newton-seconds. Newton-seconds may be converted to pound-seconds by dividing by 4.45 (Newton-seconds divided by 4.45 = pound-seconds).

All characteristics of an engine are important in selecting the proper engine to launch a specific model rocket on a specific flight. The height the rocket can reach is largely determined by the total impulse of the engine. The rocket's total weight helps determine which engine may be used for safe flights. (See Figure 1–8.)

The model rocket engine's nozzle is made of a clay material. The various shapes are modifications of the De Laval nozzle, as shown in Figure 1–9.

Specific Impulse – 80-83 Lb.-Sec. per Lb.
Exhaust Velocity – 2550-2650 Ft./Sec.

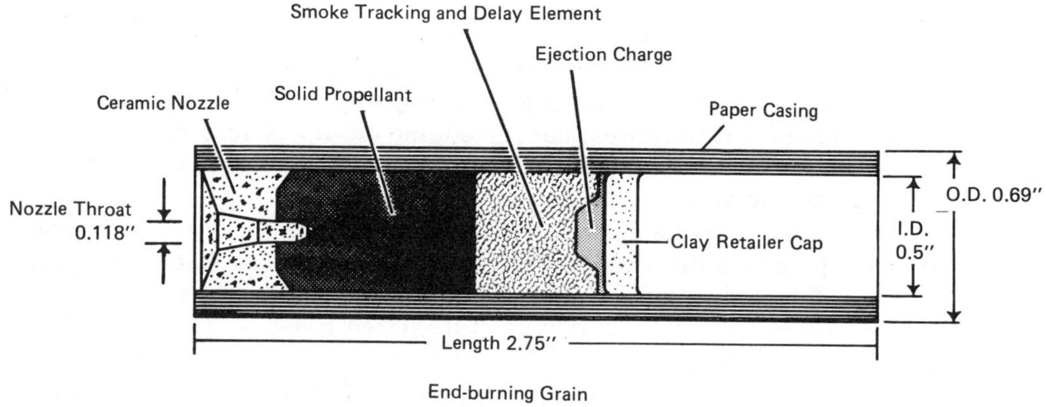

Figure 1-8. Typical model rocket engine. Typical time/thrust curve of B6-4 Series I.

LAUNCH SYSTEMS

★ A model rocket launch system consists of two elements—the launch pad and the launch controller.

LAUNCH PAD

The launch pad (see Figure 1–10) holds the rocket securely in position before launch and points the rocket in the proper direction.

The blast or jet deflector deflects the hot exhaust gases of the rocket's engine to the side. This prevents the hot gases from possibly striking dry grass on the ground and starting a fire. Also, the farther the hot gases move from the nozzle, the more they can expand and cool. The mixing of the hot gases with the surrounding air also helps cool the gases.

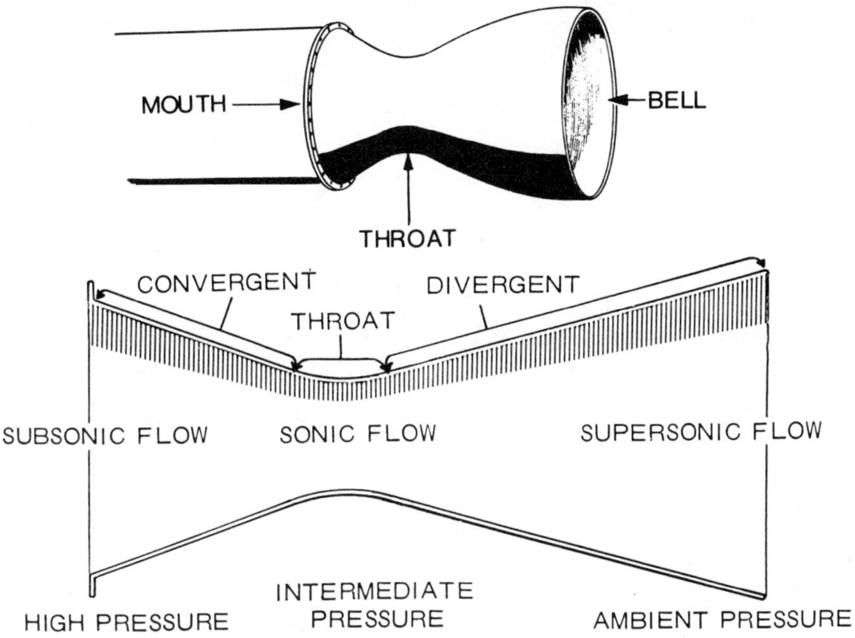

Figure 1-9. Convergent-divergent nozzle (De Laval Nozzle).

Figure 1-10. Portapad.

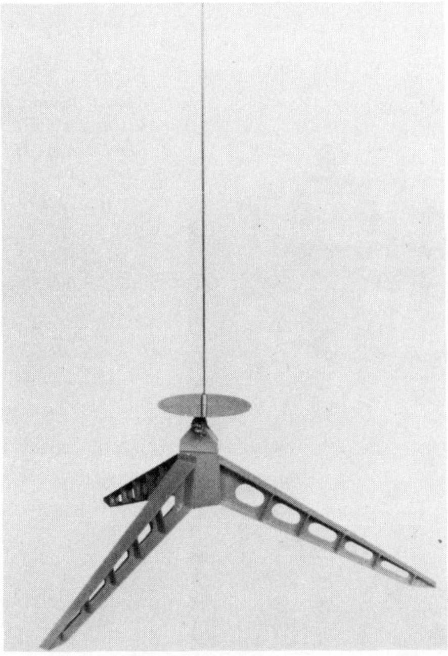

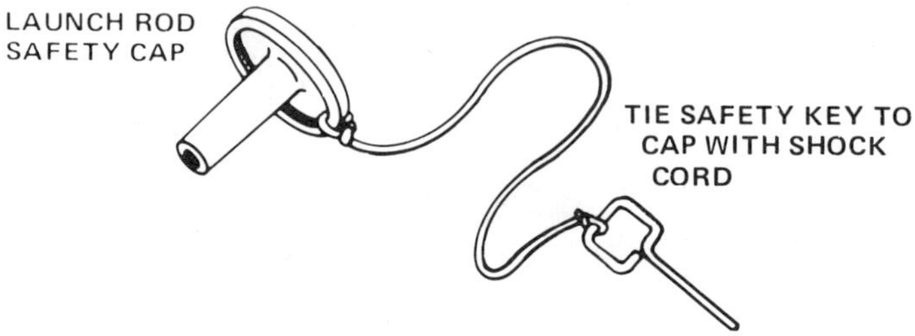

Figure 1-11. Safety cap with safety key.

The safety cap (see Figure 1–11) makes the tip of the launch rod very visible so that you can avoid it. The safety cap also covers the end of the rod to prevent you from accidentally hurting yourself with it. The safety cap has an attachment point for securing the safety key. This keeps the safety key from being lost.

The launch rod provides directional guidance for the model rocket during its first few feet of flight. By the time the model rocket leaves the launch rod (see Figure 1–12), the fins are acting against the air the rocket is moving through to keep the rocket moving in the same direction in which it was launched. The usual launch rod is 1/8 inch in diameter. The rocket's launch lug slips over the launch rod. The launch rod guides the first few feet of the rocket's flight as the launch lug slides up the launch rod until the rocket is moving fast enough for the fins to guide it. An additional section, 18 inches long, is available to lengthen the launch rod.

Figure 1-12. Model rocket leaving the launch pad. (Photograph by Michael A. Banks.)

LAUNCH CONTROLLER

★ A launch controller provides the electricity to heat the igniter to ignite the engine. The launch controller also provides positive control of the flow of electricity.

Electrical power for model rocket launches is usually provided by a 6-volt or 12-volt battery or a combination of AA or D alkaline dry cells.

A safety interlock key is provided to prevent accidental ignition. No current can flow from the power supply to the igniter until the safety key is inserted. (See Figure 1–13.) When the safety key is inserted (see Figure 1–14), a small amount of electricity can flow from the power supply through the control unit, the lead wires, and the igniter. However, the high resistance of the electrical bulb severely limits the amperage of the electricity that can flow, so the igniter does not get hot.

When the launch button (a spring-loaded switch) is pressed, the continuity light is bypassed so more current (amperage) can flow to the igniter to heat it. This places a heavy drain on the dry cells, or battery, in the power supply, but it rapidly heats the igniter to cause fast engine ignition. (See Figure 1–15.)

The Model Rocketry Safety Code requires that no one be closer than 15 feet to a model rocket being launched. Often the distance between the launch pad and those present at the launching is more than 15 feet, es-

Figure 1-13. Solar launch controller without safety key in place.

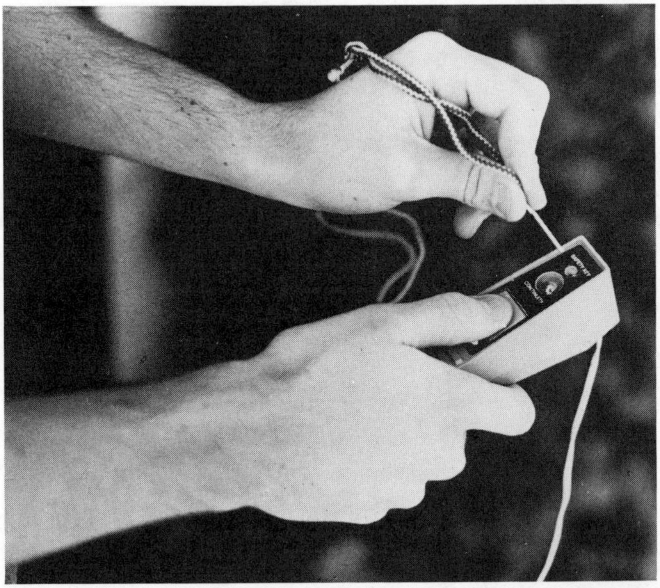

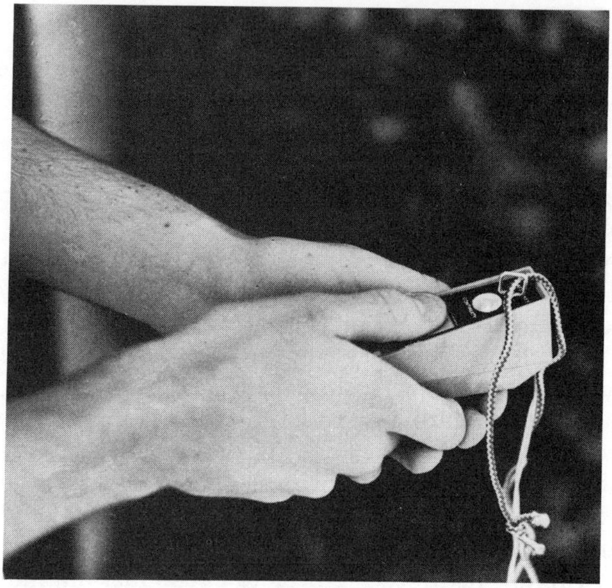

Figure 1-14. Solar launch controller with safety key inserted and continuity light illuminated.

Figure 1-15. Solar launch controller with safety key inserted, launch button being pushed, and continuity light out.

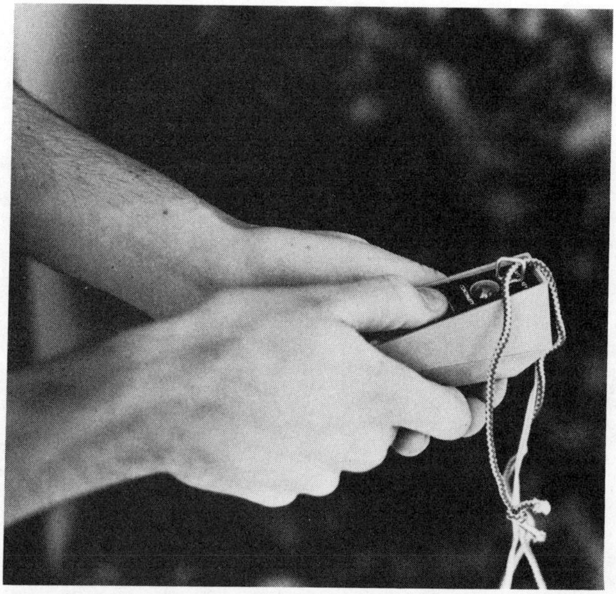

pecially when a class, club, or large group is launching. The longer the wire leads between the power supply, control panel, and launch pad, the greater the electrical resistance in the circuit. This means that launch control systems with long leads require a strong power supply.

The most common problem encountered in launching model rockets is the *misfire*. This problem may be responsible for 90 percent of all ignition failures. The problem is especially prevalent with beginners. It's usually caused by the igniter not being in direct contact with the propellant when the igniter burns.

The igniter (see Figure 1–16) must be fully inserted into the engine's nozzle. The igniter must *touch* the propellant at the base of the nozzle. Once the igniter is properly inserted, it must be secured *firmly* in place. This may be done by tightly pressing masking tape into place over the igniter. The tape should be so firmly pressed that the igniter leads are clearly "molded" into the tape. Some rocketeers press two additional pieces of tape into place to further secure the igniter. See Figure 1–17.

Another method of securing the igniter is to take a small piece of recovery wadding and roll it into a small ball. This ball is then used as a cork to secure the igniter.

One test that can be used to see if the igniter is properly secured is to hold the engine by one of the igniter leads. *Gently* shake the engine by that lead. If there is no slipping of the igniter, then the engine *probably* has the igniter firmly secured. If the tape was applied when the igniter was in the proper position, the igniter should successfully launch the engine.

Use *alkaline* dry cells in your launch system. One set of alkaline dry cells should be adequate for about a year's launches. Using regular carbon-zinc dry cells may provide electrical power for one or two launches, but rarely for more. Use of heavy-duty dry cells may provide power for about half a dozen launches.

Figure 1-16. Solar igniter. Figure 1-17 (right) *Solar igniter taped into engine.*

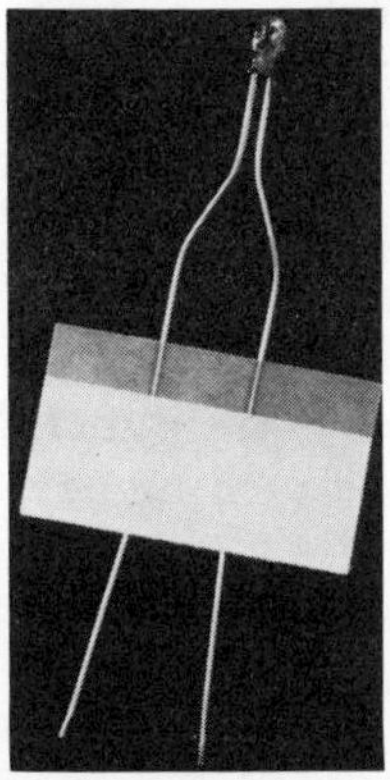

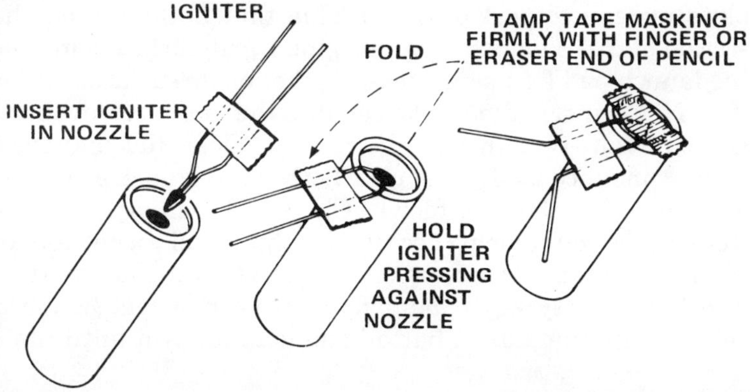

Figure 1-18. Schematic diagram of typical launch control system using external power supply.

A car or motorcycle battery may be used as the power supply. However, use extreme care when using a car battery because of the possibility of electrical shock. It is not necessary to remove the car battery from the car when using it to power your launch control system. However, it may be necessary to remove the battery from the car if you cannot conveniently get the car onto the edge of the launch field. Another problem that you may be encounter when using the car battery is that it can provide too much current (amperage). If it does, it may vaporize the filament wire in the solar igniter so fast that the thermo-plastic coating of the igniter does not heat sufficiently to burn. If this happens, the coating prevents the heat from the igniter from heating the propellant enough to ignite it. While this type of misfire is rare, it has occurred.

A number of model rocket launch-control systems are available. The most popular of these is the Estes Solar Launch Controller. This system uses four AA alkaline dry cells as the power supply. You can easily build your own launch-control system. The following schematic diagram, shown in Figure 1–18, shows the circuit of a typical launch-control system.

The microclips must be clean to make proper electrical contact with the leads of the igniter. A good rule is to clean the jaws (inside surfaces) of the microclips at least every four launches. A small sheet of extrafine sandpaper folded with the grit side outside is great for cleaning the jaws. If sandpaper is not available, the jaws can at least be scraped on the blast plate to remove part of the residue that builds up on the jaws. A small section of this sandpaper can be glued, grit surface out, to one of the legs of the launch pad for use in cleaning the inside surfaces of the microclip jaws.

Never insert the safety key into the launch controller until just before launch. Make certain that the range is clear and that the launch is imminent. Announce loudly and clearly that the launch control system is armed, and then give the countdown. Check off each of the necessary prelaunch steps on the countdown checklist for the model rocket you are launching per the instructions. At minus six, make certain that the range is clear and insert the safety key. Count down from five *loudly* and *clearly*. At zero, gently press the launch button and hold it down until the rocket takes off.

Do not press hard on the launch button. The only function of the launch button is to act as an electrical switch making contact to "short out" the continuity bulb. This permits the full current from the power supply to flow to the igniter. The resistance of the nichrome wire in the igniter to the current flow heats the igniter to cause engine ignition and liftoff. Pushing too hard on the launch button can cause the electrical contacts to jam. This will cause each rocket connected to the launch controller to take off as soon as the safety key is inserted until the jammed switch is repaired.

LAUNCH SITES

★ The basic requirement for a model rocket launch site is plenty of open space.

For small rockets, using 1/2A, A, or B engines, a school football field or baseball field will be sufficient. The field should be clear of obstacles (trees, benches, and so on) and, ideally, should have no large trees growing nearby. This cuts down on the hazard of losing a rocket if it drifts out of the field. The ideal field should be as far as possible from housing, roads, and power lines, and away from any air traffic.

Always secure the owner's permission before launching from private property.

The rule of thumb for selecting a launch site (or for scaling model rocket activities) is to launch in a field the shortest dimension of which is at least one-fourth the expected altitude of the rockets to be flown. This should provide sufficient recovery area with reasonable wind conditions, but the bigger the field, the better!

The launch site should be easily accessible to those taking part in the launch. It should not be near tall buildings, electrical or telephone lines, airport flight paths, highways, or anything else that could present a hazard to those launching or for whom the launchings could constitute a hazard. The launch site should not contain any easy-to-burn material, such as dry grass or weeds. The area selected for launch and recovery should be large enough to ensure that all flights can be safely made and recovered within the bounds of the selected area. The area should be divided before the launch starts into launch pad area, rocket preparation area, spectator area, parking area (if needed), and recovery area. (Figure 1–19 shows a model rocket launch site with the different areas marked off.)

The size of the area needed will be determined by the types and weights of the rockets to be flown, the types of engines to be used, and the wind. By careful aiming of the launch rods and careful selection of engines, even a fairly small area may be safely used, as shown in Figure 1–19.

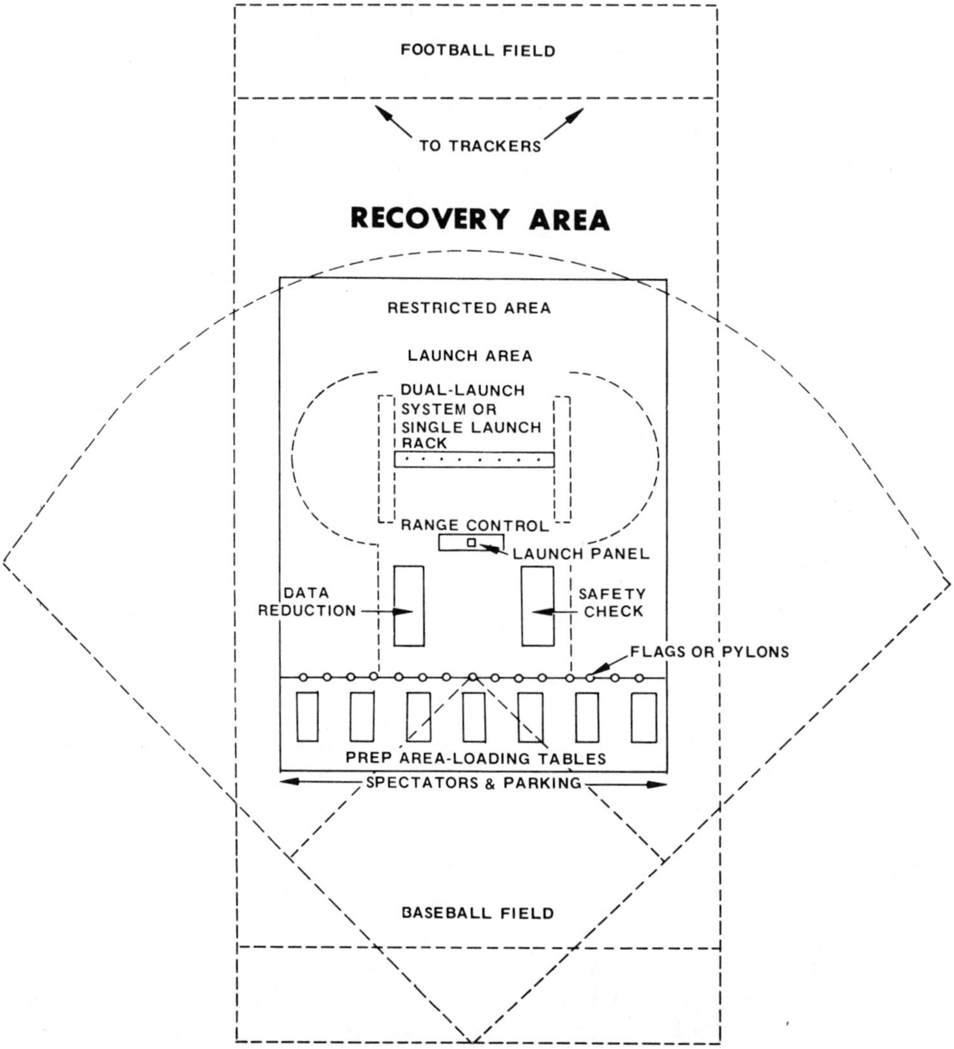

Figure 1-19. Model rocket launch site showing possible placement on various open areas.

REGULATIONS

★ Always follow the National Association of Rocketry/Hobby Industry of America Model Rocketry Safety Code.

Be sure that you have secured the necessary permits if your area has special regulations on launching model rockets and/or using a public area such as a park, a school playground, or parking lot.

The National Fire Protection Association has enacted a special code to encourage safe model rocketry and to discourage unsafe amateur rocketry. This special code is NFPA Code No. 41L, "Code For Model Rocketry 1968." Following the code specifications is common sense, because mixing propellant is not only dangerous but illegal.

The Federal Aviation Agency has enacted special regulations pertaining to model rockets as "Part 101—Moored Balloons, Kites, Unmanned Rockets [and Unmanned Free Balloons] [New]." This regulation sets strict standards for the launch of unmanned rockets but exempts model rockets that follow the Model Rocketry Safety Code from the regulations governing the operation of unmanned rockets.

Model rocket engines can be shipped by U. S. mail, further illustrating their safety.

Model rocketry is an entire technology in miniature, as this chapter and the rest of this book illustrate, and the only true limitations to model rocketry activities are those of the imagination.

Model rocketry has proved that it is a safe hobby. The cooperative efforts of millions of individuals over more than 25 years and in well over 200 million launches have made model rocketry a safe and widely accepted hobby. Let's all do our part to keep this excellent safety record intact.

HISTORY

The first rockets were flown in northern China over 2,000 years ago. These were solid fuel rockets powered by "black powder," which is similar to gunpowder.

The first liquid fuel rocket was test-flown by Robert Goddard in 1926. (See Figure 1–20.) Goddard's work took rocketry to a new level, giving it previously unheard-of potential for moving payloads of any sort over great distances, and perhaps in space. Goddard also did work describing the requirements of interplanetary flight.

Between World War I and World War II, there was much interest and speculation in the use of rockets for space travel and for moving payloads from one point to another here on Earth. Experimental-rocket societies popped up nearly everywhere in the 1930s, and many individuals at-

Figure 1-20. Robert Goddard and world's first liquid-fuel rocket.

Figure 1-21. Apollo lift-off (Saturn V).

tempted—usually with disastrous results—such activities as sending mail via rocket and even flying onboard rockets. There were also rocket-powered cars and gliders.

The launch of the Russian Sputnik, the first artificial satellite to orbit the Earth, touched off a race that is still running. American scientists rushed to catch up with Russian developments in rocketry and surpassed the Russians on July 20, 1969, when Neil Armstrong became the first man to set foot on the moon. (See Figure 1–21.)

In a sense, model rocketry grew out of the space race. In 1954, the first model rockets were built by Orville Carlisle of Norfolk, Nebraska. Carlisle sent samples of his model rockets to G. Harry Stine, who was an engineer at White Sands Proving Grounds in New Mexico, the site of most of the rocketry experiments in the United States in the years following World War II. Stine had been contributing a monthly column on space travel and rocketry to *Mechanix Illustrated.* Stine tried out the rocket engines and the rockets. He improved them and began marketing various model rocket kits through a company he founded called Model Missiles, Incorporated.

Model rocketry is not restricted to the United States. The hobby spread quickly to other parts of the world, and many new manufacturers appeared in the United States. Today there are model rocket manufacturers in the United States, Czechoslovakia, Poland, and other countries.

Figure 1-22. Vern and Gleda Estes. (Photograph by Peter Galindez.)

Vern Estes founded Estes Industries in his garage in Denver, Colorado. His small company was so successful that he soon moved to Penrose, Colorado, the present site of Estes Industries. Vern and his wife, Gleda, (see Figure 1–22), emphasized good service and the educational uses of model rocketry as well as providing products of reliable quality, an approach that helped Estes Industries to become the leader in model rocketry. Penrose, Colorado, is indeed "the model rocket capital of the world."

2

MODEL ROCKET CONSTRUCTION TECHNIQUES

The construction of a model rocket can appear to be a fairly straightforward and simple task. It is, if you take the time to follow kit instructions and don't rush the process. Attention to detail is vital. See Figure 2–1.

The major structural elements of a model rocket are the body tube, the nose cone, the fins, the engine mount, and the recovery system. Accessory components include the launch lug, the shock cord, and the shock cord mount.

Figure 2-1. "Exploded" view of model rocket with parts labeled.

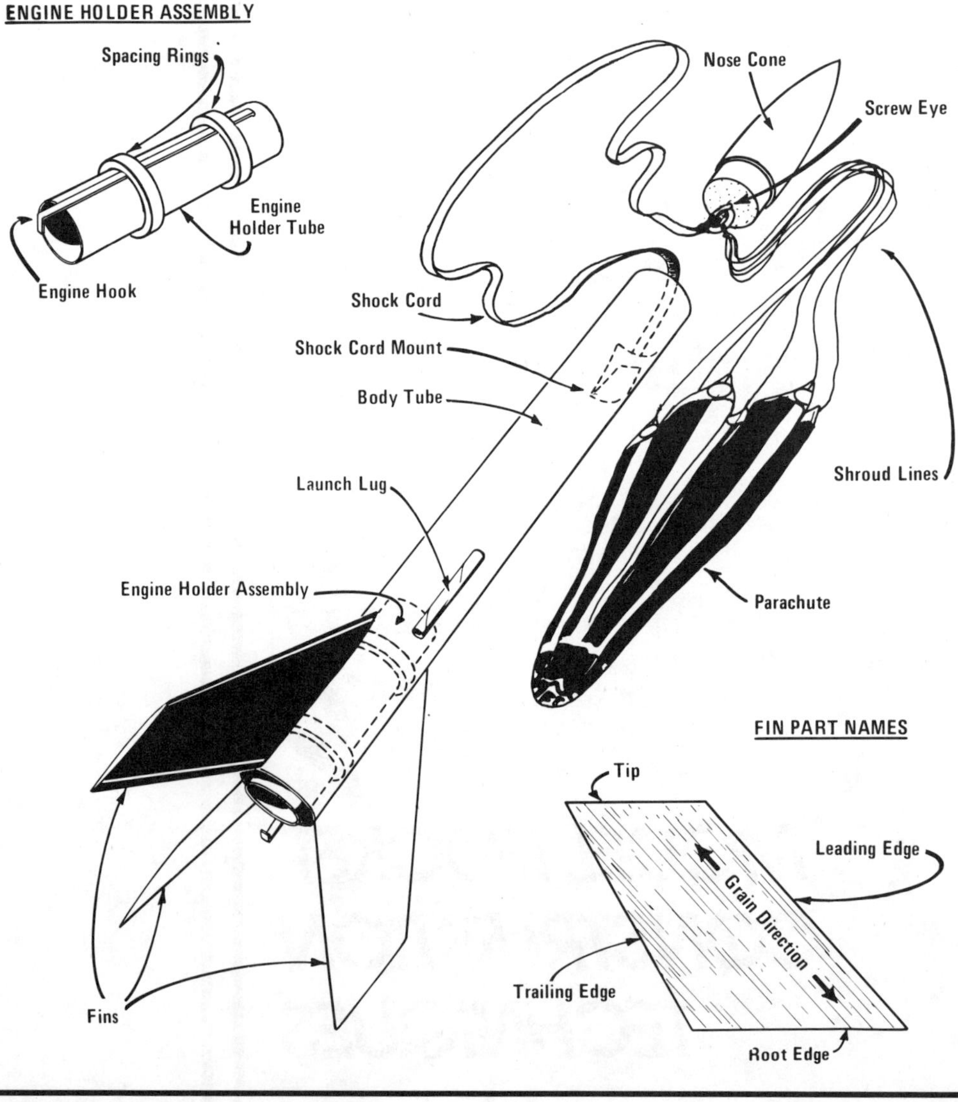

BODY TUBES

★ The body tube provides the frame that furnishes the basic shape and supports all the other parts of the rocket.

Body tubes are made of paper wound tightly in several layers. The layers of paper may be wound in a spiral or in a parallel pattern. The spiral pattern is used most often because it provides a more uniform circular cross section.

FINS

★ Fins provide the in-flight guidance for model rockets.

Fins can be made of balsa wood, basswood, plywood, plastic, or cardboard. See Figure 2–2A. Fin thickness can vary from 1/32 inch to 1/4 inch. Balsa is an extremely lightweight and reasonably strong wood. Basswood, a somewhat heavier, denser wood, is sometimes used when more fin strength is required. Thin plywood of the type used to build model airplanes can also be used as fin material, but it is usually used only with high-power designs where fin strength is a vital element. Although plywood is much stronger than either balsa or basswood, it is correspondingly heavier, and this affects model rocket performance.

Balsa is very porous, with large grain structure, and therefore requires filling with sanding sealer or some other suitable substance before it can be sanded to a smooth finish for painting. Basswood requires less filling and sanding than balsa because it is a harder wood. Plywood requires the least finish preparation because it is usually prefilled and sanded.

Figure 2-2A.
Fin.

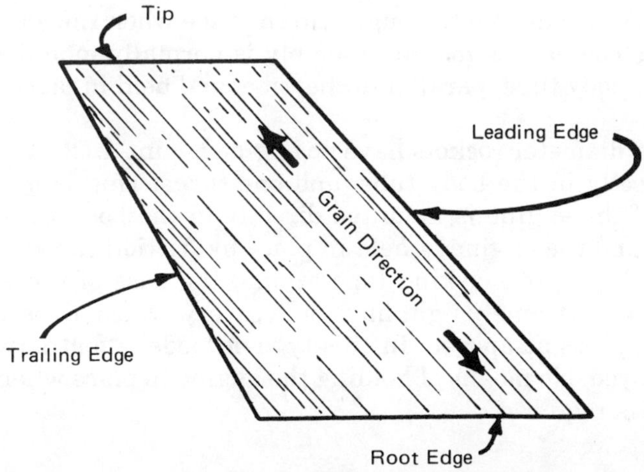

Balsa is classified by grades based on density. Careful selection of balsa for the proper grade and grain direction can be important, especially for the wings of gliders.

NOSE CONES

★ Nose cones have a great effect on the drag experienced by model rockets.

Nose cones are usually made of balsa or plastic. A few high-power model rockets use pine or hardwood for nose cones. Plastic nose cones are hollow and may be made by injection-molding or blow-molding. Some nose cones are made in two sections that must be glued together before use.

A nose cone has an attachment point for securing it to the shock cord. Most plastic nose cones have an attachment loop molded into them. Balsa nose cones are usually adapted for attaching parachutes by twisting a screw eye that is anchored in glue into their bases. Sometimes a nose cone will have weight, in the form of clay, added to it to move the center of gravity of the model rocket forward.

ENGINE MOUNTS

★ An engine mount holds the model rocket engine in the proper position during thrust, coasting, and recovery. (See Figure 2–2B.)

The engine mount consists of a cardboard tube to hold the engine in place. There is often a cardboard ring known as a "thrust ring" at the forward end of the tube. The engine mount usually includes an "engine hook," a metal clip to hold the engine in the tube when the engine's ejection charge fires. The engine mount assembly is normally mounted within the model rocket body tube, parallel to the tube, and held in place by centering rings.

Smaller diameter rockets have no engine mount. Instead, the engine is inserted directly in the body tube, and one thrust ring is mounted in the body tube. If the engine is mounted directly in the body tube, there is no engine clip, and the engine is held in place by friction caused by wrapping one or more layers of masking tape around the base of the engine. Some models that are extremely light have no recovery system to be activated and eject their engines at apogee. In this type of model, "featherweight recovery" is employed. No means of holding the engine in place when the ejection charge fires is required.

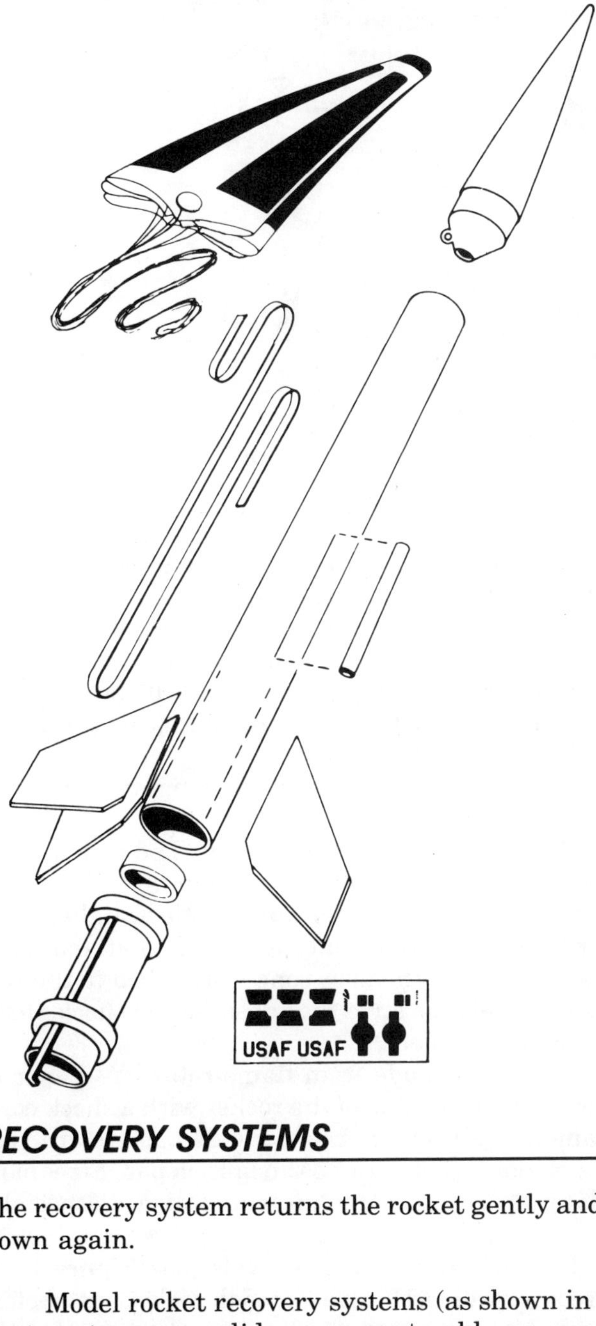

Figure 2-2B.
Engine mount with thrust ring, centering rings, and engine hook.

RECOVERY SYSTEMS

★ The recovery system returns the rocket gently and safely to the ground to be flown again.

Model rocket recovery systems (as shown in Figure 2–3) include parachutes, streamers, glide recovery, tumble recovery, helicopter, break-apart, and featherweight recovery.

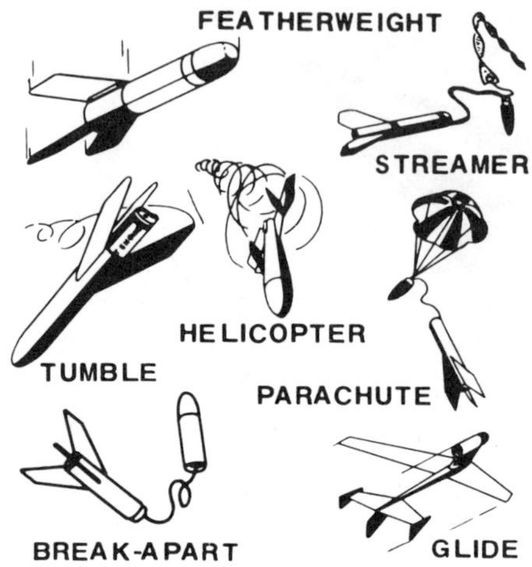

Figure 2-3. Recovery systems.

Tumble recovery is initiated indirectly by the engine ejection charge. The engine is either ejected, which upsets the stability of the rocket, or the engine is moved rearward by the force of the ejection charge. This also causes loss of the rockets stability. In each case, the rocket tumbles on its way to the ground, creating high drag, which slows its fall. If the engine is not ejected, a means must be provided to hold it in place, but in a new location. Usually, an extralong engine hook is used.

Featherweight recovery is used only with very small, lightweight rockets that cannot be damaged and cannot damage other objects on impact. The engine is ejected at apogee and falls back to the ground. The rocket itself also falls freely, but it is so light that it is not damaged on impact.

Parachutes (see Figure 2–4) are made from thin plastic or lightweight silk or nylon fabric. Cotton or nylon shroud lines are used to attach parachutes to the nose cone of a rocket, and the nose cone is attached to the rest of the rocket via a shock cord (an elastic strip that absorbs the shock of the ejection charge and parachute deployment).

Streamers (see Figure 2–5) are made from flame-retardant paper or thin Mylar and are attached to the nose cone of the rocket with a shock cord. Both parachutes and streamers are high-drag devices. Streamers are very useful in safely returning a model rocket close to the launch pad. Streamers do not produce as much lateral drift as parachutes and so are especially useful on windy days.

Attach streamers and parachutes to snap swivels (see Figure 2–6). Then attach the proper recovery device to the model rocket just before launch. This permits choosing the proper parachute or streamer for the existing wind conditions just before each launch. Also, this method permits the rapid change of parachutes in case a parachute is damaged.

Figure 2-4. Model rocket descending under parachute.

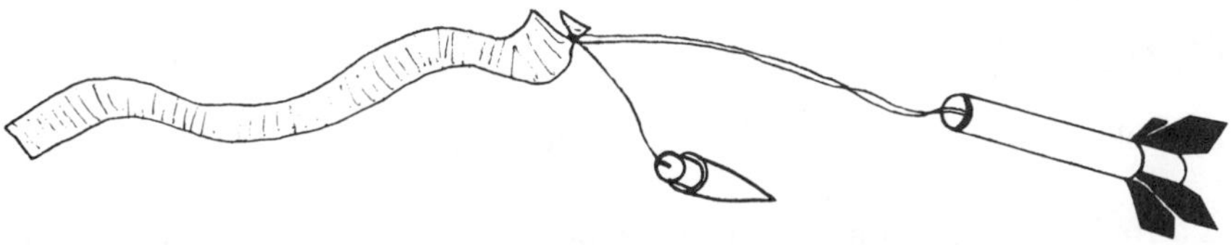

Figure 2-5. Rocket descending under streamer.

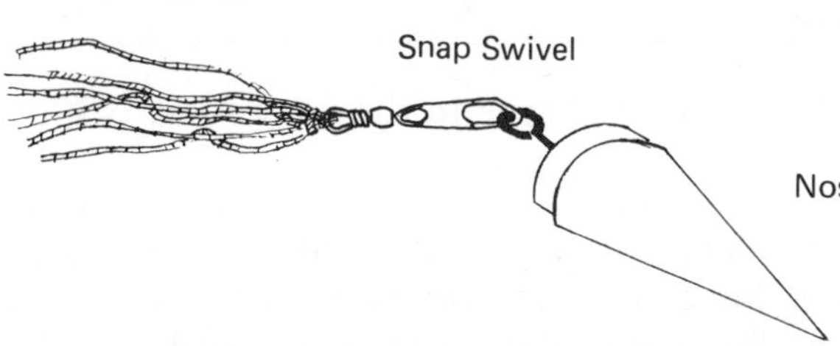

Figure 2-6. Snap swivel attached to parachute.

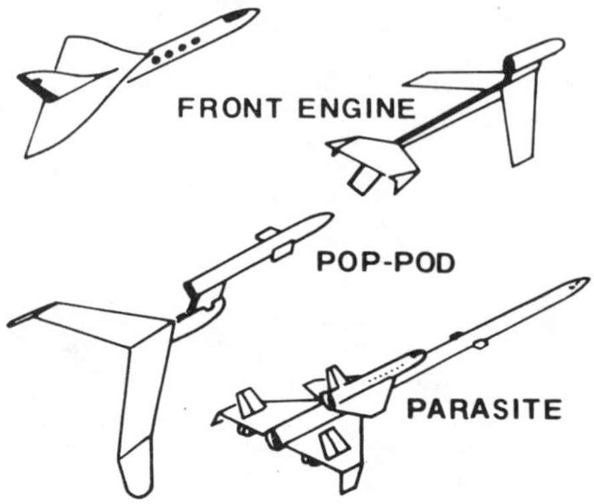

Figure 2-7. Rocket-launched gliders.

Glide-recovery devices (see Figure 2–7) include wings and para-wings. Some glide-recovery systems are deployed from within the rocket's fuselage or consist of wings that are folded along the sides of the fuselage and pop out upon engine ejection. Some models have fixed wings, and engines mounted on long booms at their forward end. These model rockets, known as "boost gliders" or "rocket gliders," change their centers of gravity when the engine and its mounting pod are ejected. The configuration changes from that of a rocket to that of a glider. The basic types of boost gliders are front engine, rear engine, and pop-pod.

LAUNCH LUGS

★ The launch lug guides the model rocket until it leaves the launch rod.

Launch lugs are small tubes of paper that are glued to the side of the rocket parallel to the rocket's long axis.

SHOCK CORDS

★ The shock cord connects the nose cone to the body tube.

The shock cord is essentially a long rubber band that secures the nose cone to the body tube. Its elasticity "forgives" if you accidentally choose an engine of the wrong delay time. If the recovery system is ejected while the rocket is still coasting fairly fast, the shock cord stretches to keep the nose cone and recovery system attached to the rest of the rocket.

SHOCK CORD MOUNTS

★ The shock cord mount secures the shock cord to the body tube.

The shock cord mount secures the rubber shock cord to the paper body tube. This attachment must be strong enough to prevent the shock cord from being detached from the rocket, even if the stress it experiences at ejection is quite strong.

GLUES

★ Glue holds the various parts of the model rocket securely in place.

There are several types of glues that can be used in building model rockets. For most rockets, ordinary "white glue" (such as Elmer's or Wilhold) is fine. White glue is a water-based adhesive, so it is easy to clean up while it is wet. It is nontoxic and nonflammable. It is designed to be used with wood and paper, the major elements of model rockets. It dries clear and is inexpensive. The only disadvantages of working with white glue are its relatively long setup time and the time required for it to dry to full strength (up to 70 hours).

Yellow glue is similar to white glue, but it is specially formulated for gluing wood. It penetrates wood better than white glue and forms stronger joints. Yellow glue also sets quicker than white glue, which means that parts being glued don't have to be held in place as long. When fully dry, yellow glues can also be sanded with greater ease. Yellow glues are more heat- and solvent-resistant than white glues. Brand names of yellow glues include Titebond, Weldwood Wood Worker Glue, and Elmer's Wood Glue.

Epoxy is a very strong and versatile adhesive. It works well with most metals, wood, glass, some plastics, and paper. Most epoxies can tolerate temperatures up to 200 degrees Fahrenheit and are resistant to most solvents. The major advantage of epoxy over white or yellow glues is its strength. The disadvantages include price, the requirement that epoxies be mixed before using, its dangerous fumes when freshly mixed and while curing, its weight, rigidity, and the fact that epoxy glue does not adhere to body tube exteriors and balsa as well as white or yellow glues.

When working with plastic parts (fins or components of the AstroCam, which will be discussed later, for example), you will want to use plastic glue. There are two types of plastic glue suitable for use in model building: liquid plastic cement (solvent) and "tube-type" plastic cement. Liquid plastic cement is excellent for working with small or precision parts because it actually fuses plastic pieces together. Tube-type plastic cement, which comes in a tube and is rather thick, is good for most applications. Care should be taken not to use too much of either type, because plastic tends to "melt" when too much cement is used. Also, the fumes of some types of plastic cement are toxic.

Cyanoacrylates ("magic glues") can glue most nonporous materials. Since balsa and cardboard are porous, cyanoacrylates have limited uses in model rocketry. In addition, cyanoacrylates can glue skin to skin or to other materials. This makes them dangerous to use. While fingernail polish remover will usually dissolve cyanoacrylates, extreme care must be observed when using them.

There is a type of cyanoacrylate that will bond balsa and cardboard. Use it *exactly* as instructed. It is great for initially gluing fins to body tubes. Always follow such bonds with a strong fillet of white glue rubbed well into the balsa and the cardboard. Use care whenever using cyanoacrylates.

ASSEMBLING A MODEL ROCKET KIT

WORK SURFACES

★ Always work on a surface that is firm, smooth, and can be easily repaired or replaced in case of accidental damage.

When you are assembling model rockets in a classroom situation, a primary consideration should be protecting work surfaces. Ideally, you will be working on surfaces that are either hard to damage or will not create problems if they are damaged. Since this is normally not the case, you should cover the work surfaces used by students with a material that will be cut-resistant and glue-resistant. Several layers of newspaper will serve to protect tabletops from scratches and dents, and will absorb glue and paint. A section of corrugated cardboard, such as a foot-square section cut from a box, makes an excellent work surface.

IDENTIFY ALL PARTS

★ The first step in assembling a model rocket kit is to identify each part using the construction diagram that comes with the kit.

Normally, a kit will include a step-by-step instruction sheet that will include an "exploded" diagram of the completed model rocket and a parts list. It is best to lay out the parts of the rocket and find each one on the diagram, as shown in Figure 2–8, comparing the parts with what is on the parts list. Make certain that all parts are present. Become familiar with each part's name and its place in the rocket. Then you can begin construction.

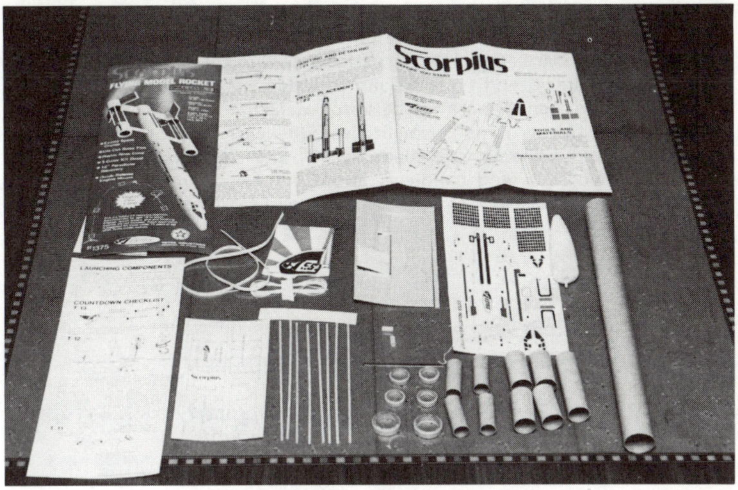

Figure 2-8. Parts spread out on table with instruction sheet. (Photograph by J. Kevin Wolfe, ARTifacts.)

NECESSARY TOOLS

You should have, in addition to the kit and the instruction sheet, a pair of scissors, an X-acto-type knife, a ruler with a metal edge, masking tape, fine sandpaper, a pencil, and glue. You may also wish to have a razor saw for cutting thick balsa and/or plastic. You will need a cutting board to protect your work surface. Your work area should be large enough to allow all the kit's parts to be laid out. The work area should be well lighted and well ventilated.

Some kits require special supplies, such as tube-type plastic cement.

READ THE INSTRUCTIONS

★ After familiarizing yourself with the parts of the kit, *read the instructions for the kit.*

No matter how "simple" a kit appears, there will almost always be some important detail that you will miss if you don't read the instructions. After reading through the instructions, perform a "dry run" of the assembly by fitting together the parts without gluing them.

CUT OUT FINS

If you have to cut your rocket's fins from a sheet of balsa, make certain that the fins are cut out in such a way that the leading edge of each fin is parallel to the grain of the balsa. (See Figure 2–9.) The root edge (the edge that

CUTTING

Draw a full-size fin pattern on stiff paper or cardboard. Cut out the pattern, position it on the fin stock, and trace around it with a pencil or ball point pen to mark the balsa for each fin.

Use a metal straightedge whenever possible. Hold knife or saw blade at 90° angle to surface being cut, and handle at about 45° for clean cut. If blade is dull or held too high balsa tends to tear.

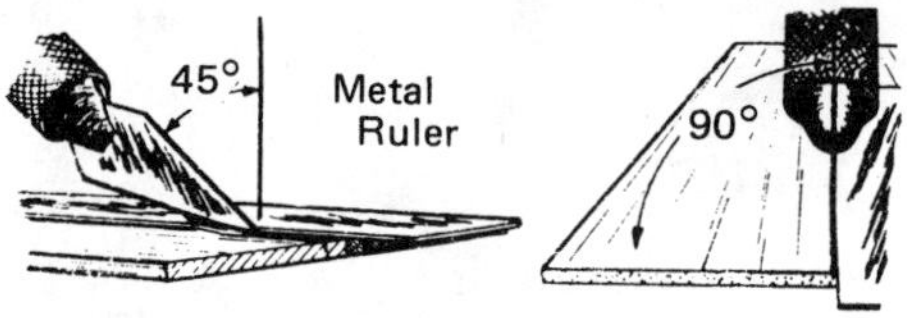

Figure 2-9. Balsa fins being cut out properly using metal-edge ruler and modeling knife.

attaches to the body tube) of each fin is usually perpendicular to the grain of the balsa. Cutting out fins in this manner produces strong fins. If you don't cut the fins out in this way, they are very likely to snap off when stress is applied to them.

When using a modeling knife to cut balsa (or any other type of wood), don't rush. Using a ruler and a pencil, carefully mark the lines you will cut. After marking the lines, run the edge of the knife lightly along the lines, scoring a path for the blade to follow. To make a straight, clean cut, place a metal ruler or metal-edged ruler over the fin. Use the metal edge as a guide and to prevent accidentally cutting into the fin. Make your final cuts using firm, even pressure as you draw the knife steadily along the line. Try to cut completely through the balsa.

CUT OUT OTHER PARTS

Razor saws are most useful in cutting apart sections that are made of blow-molded plastic. Razor saws are also handy when cutting balsa blocks, such as when roughing out nose cones.

If you are using a razor saw, work slowly. Move the saw with short, even strokes. When using a modeling knife or saw, do not cut toward your body. Instead, make your cuts parallel to yourself, moving the balsa sheet as necessary to make cuts in different directions on the sheet. If you are right-handed, cut left to right. If you are left-handed, cut right to left.

Paper can be cut with scissors or, if you have a steady hand, with a modeling knife. When cutting paper with a modeling knife, you should hold the paper firmly in place on a flat surface with one hand while cutting in a direction *away* from the hand holding the paper. As when cutting balsa, you should use a continuous, light-but-firm motion, parallel to your body.

GLUING

When gluing balsa and paper, first put a line of glue on one of the parts (such as the root edge of a fin) and then touch it to the other part. Separate the parts. Rub the glue firmly into the balsa grain on the root edge of the fin. Allow the glue a few seconds to "set" or become "tacky." Then apply some more glue to the first part and press the parts together, lightly but firmly. You will have to hold the parts together until the glue bond becomes strong enough to hold the parts without moving. This normally requires less than a minute, but you may wish to lightly attach the pieces together with masking tape to ensure that they will not move while the glue is drying.

For strength, all glue joints should be reinforced. This second glue layer is called a "fillet." The fillet should be smoothed out with a finger. White glue dries and sets primarily by the evaporation of water. As the first layer of glue dries, it turns from white to clear. After the first glue layer has dried, apply the second layer or fillet. Rub this layer up onto the balsa and onto the body tube. When the fillet layer has dried, it is clear and very strong. In fact, this glue layer may be stronger than either the balsa or the cardboard. The fillet layer should extend onto the balsa and onto the body tube. This second layer acts to "weld" the parts together.

PARACHUTES

Parachutes are normally shaped as hexagons or octagons. In kits, you will be supplied with square pieces of plastic parachute material with the parachute shape silk-screened or printed on it. After cutting the parachute from the plastic square, you should lay the parachute out flat, printed side up, and attach the shroud lines. In the case of an octagon-shaped parachute, you should cut four equal lengths of the shroud line material (string) from the supply included. Each shroud line should be twice the diameter of the parachute. Attach the ends of each shroud line to adjacent corners of the parachute, using the tape discs supplied with the kit. With a hexagonal parachute, you should cut the string into three equal lengths, each twice the diameter of the parachute, and attach these in the same way.

Tie a knot in each end of a shroud line. Make a large loop in the string and carefully place the loop inside a tape disc. Place the tape disc carefully on the marked place on the parachute corner. Rub the disc down *firmly* until the disc and the plastic of the parachute are firmly molded around the shroud line. Repeat this procedure for the other shroud line ends. (See Figure 2–10.)

PLASTIC PARTS

Carefully cut away any "flash" (excess plastic) from all plastic parts and sand, if necessary, with extrafine sandpaper to smooth them (see Figure 2–

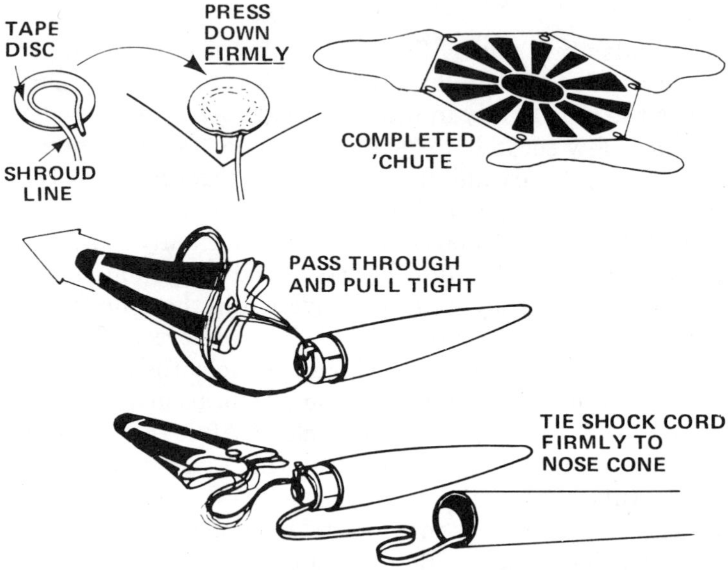

Figure 2-10. Shroud line attachment.

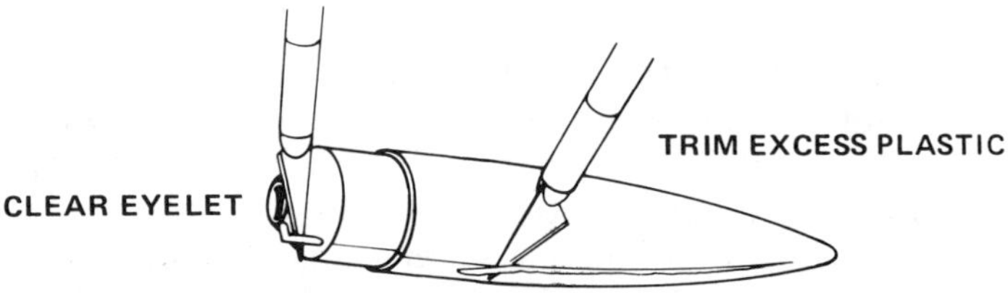

Figure 2-11. Cutting away flash from a plastic part.

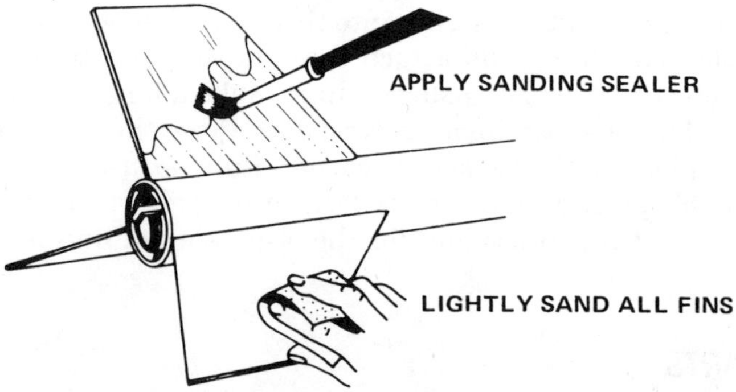

Figure 2-12. Balsa filling and sealing.

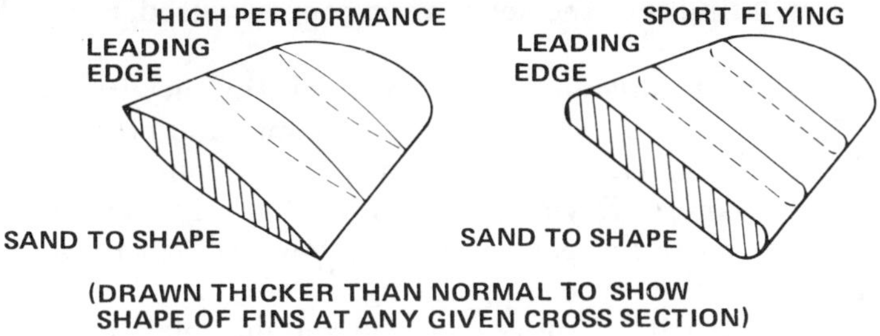

Figure 2-13. Rounded and air-foiled fins.

11.) Thoroughly wash all plastic parts with detergent and fully rinse. This is to remove the mold release oil and fingertip oil from the plastic parts.

BALSA FILLING AND SEALING

When finishing a model rocket for painting, fill in the rough balsa surfaces with some type of filler coat and sand the surfaces as smooth as possible, as shown in Figure 2–12. Apply a coat of balsa filler, let dry, and then sand with extrafine sandpaper. Repeat two or three more times, as necessary, until the surface looks and feels smooth. You may also consider lightly coating and sanding body tubes to fill in the spiral windings on their surfaces.

Placing fins exactly parallel to the rocket's axis is important. Improperly placed fins may cause the rocket to spin in flight, robbing it of energy that could be used to achieve greater heights.

Model rockets may be flown with their fins "square-cornered." Rounding the corners of the fins will improve the performance of the rocket. Sanding the fins to an airfoil shape will increase the performance of the model rocket even more. See Figure 2–13.

Fins may be reshaped from their recommended shape, but models with fins not made according to the instructions *must* be stability-tested before their first flight.

PAINTING

Painting a model rocket involves some preparation. The wooden parts of the model (usually fins and nose cone) should be sanded, filled with balsa sanding sealer or dope, and sanded again when the sealer dries. This should be repeated until a smooth, plastic-like surface is achieved. Use progressively finer grades of sandpaper as you work if you want a perfect finish. Paper and cardboard parts should be dusted off. Oily fingerprints should be wiped away. Most surfaces will have fingerprints, even if you cannot see them.

Enamel paint is less expensive and easier to work with than butyrate dope, but you will find that the dope, when properly used, can yield a better finish.

By far, the most convenient method of painting is with common spray cans. (See Figure 2–14.) A wide variety of enamel and other types of paint are available in spray cans, ranging in quality from Testor's or Pactra brands of enamel paint to Krylon paint, which has a fine texture and results in a very smooth finish. For ease of use and low cost, spray-can painting is recommended over other types of painting.

When painting with a spray can, you should hold the can stationary and move the model. A rolled-up piece of newspaper, inserted in the engine mount of the model, can serve as a handle to hold the model. When painting, hold the model about 12 inches from the can and move the model slowly back and forth in front of the spray. If this is awkward, you can hold the model stationary and move the spray can if you always keep the spray pointed at right angles to the rocket's axis. Do not rotate the can by "aiming" it at the rocket as you spray.

Do not stop moving the model, because excess paint will rapidly build up and create runs or drips. Start the spray *before* the rocket is in the path of the spray and keep the spray on until the rocket is *past* the spray.

Spray a light coat of paint on the model first and allow this coat to dry. Follow with a second coat, a third coat, and a fourth coat, if necessary, to cover the model completely and smoothly. (See Figure 2–15.)

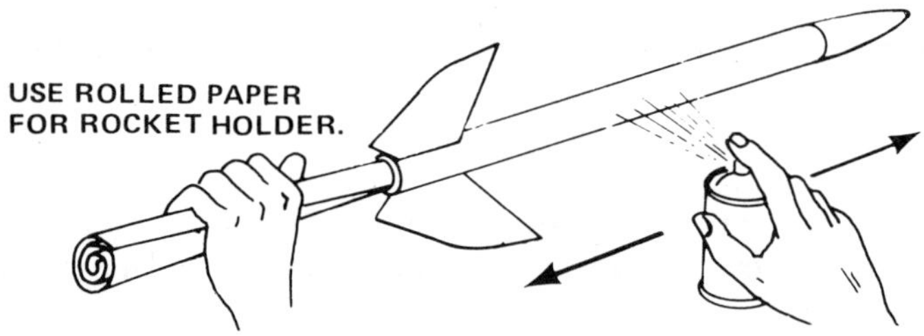

Figure 2-14. Model being sprayed with spray paint.

Figure 2-15. Masking and painting second color.

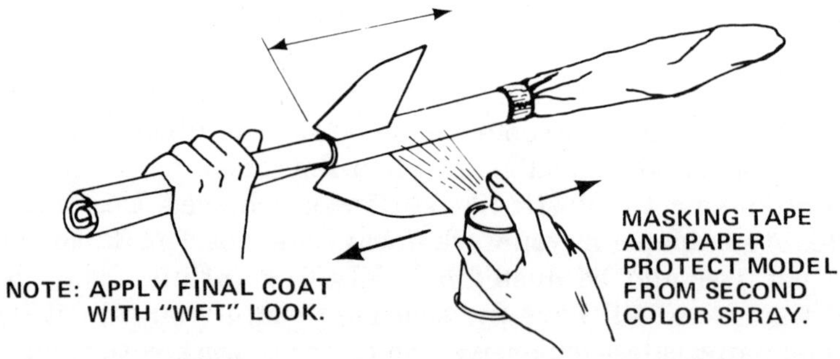

If you are painting a model of two or more different colors, you will have to mask off the various sections. It's a good idea to first paint the entire model with the color that will cover most of the model. After that color is dry, mask the portion of the model that is to remain that color. This method of applying the color with the largest area of coverage first is not recommended if the first color is dark, and a light color will be applied later.

The simplest way to mask off a section of a model is to wrap a piece of masking tape around the model at the point where you wish the original color to end. Press the tape on *lightly* in order not to pull the original paint off when you remove the tape. After doing this, tape a piece of a dry cleaner's plastic garment bag or a newspaper over the original color to protect it from overspray. Fully enclose the portion that is to remain this color to prevent "overspray" particles from drifting onto the initial color. Then paint the exposed section whatever color you want. After the new paint has dried, carefully remove the plastic or newspaper and the tape. If some of the paint comes off with the tape, touch up the damaged section using a small paintbrush dipped in paint sprayed onto a piece of cardboard from the can. Or, you can wrap colored tape around the damaged section. This technique is frequently used by modelers if it makes an attractive color scheme.

Brush-on paints are often less expensive than spray paints. They are very difficult to use to secure a nice finish. Also, good brushes are relatively expensive and must be kept clean. Few modelers can achieve really attractive finishes using brush-on paints. The edges of the brush strokes tend to apply the paint more thickly. When the brush strokes overlap, as is often necessary to fully cover the model, the problem is made worse. Thinning the paint, whether enamel or dope, is usually necessary. Several light coats of paint look better than one thicker coat.

When applying brush paints, use long, slow strokes, all in the same direction. Ideally, the brush strokes will be in the direction of the longest dimension of the model (or portion of model) being painted.

Masking off sections of a model to be painted with brushes is accomplished much the same way as when using spray paints. The only difference is that you do not have to protect the original color from overspray. Brush paints take much longer to dry completely than spray paints do. This is a disadvantage. To speed up drying time, thin the paint as much as possible, and spread the paint on in light coats. Also, be sure to have on hand several sizes of brushes. Large brushes should be used when you are painting large areas, such as fins. Small, fine-haired brushes should be used for detail work and for painting "transition" areas where masking is required.

Airbrushing is the best way to achieve a very high-quality finish, but it has the disadvantage of requiring an expensive piece of equipment, the airbrush unit. This constitutes an air compressor, a paint bottle, the spray head, and a connecting hose. Also, you frequently must thin paints and/or mix them to achieve desired colors and consistencies. The airbrush must be carefully cleaned after each color or use, and sometimes more often, to prevent clogging.

For all this expense and trouble, you have the ability to completely control the color, texture, and thickness of the finish on your models. Interesting effects, such as shading from one color to another or fading a color through various shades, can be achieved thanks to the control the spray head gives you over the width of the paint spray. Interesting designs and patterns can also be painted on a model using an airbrush.

Whatever method you use to paint your model, there are several additional things you can do to enhance the appearance and the durability of the finish. If you are painting the model a dark color, you may wish to spray on an undercoat or "primer" of a lighter color that will not "bleed through." This will give the final color a base to adhere to and will help seal any rough areas you may have missed during your painting preparations.

If you make a mistake during painting that results in a run or a lump of paint, you can correct this by letting the paint dry and then sanding off the run or lump along with an area surrounding it of about an inch. Then you can repaint the area, blending it into the original coat of paint.

DECALS

After you have finished painting your model, you will probably want to apply decals. Before applying them, you should dust off the model and carefully wipe it with a clean, oil-free, lint-free cloth to remove fingerprints. When applying the decals, you will find that special chemicals to help the decal mold better to the surface and adhere better will aid in the application (see Figure 2–16). Several brands of decal setting solution are available in hobby shops.

After applying the decals, let them dry thoroughly. You can help the process along by blotting up any excess water or decal setting solution with a paper towel. When you are sure that everything has dried completely, you can spray a coat of clear dull or clear glossy spray to protect the finish and the decals from scratches and moisture. The glossy spray coating will result in a shiny finish, whereas the dull spray coating will give you a finish that looks much as it did before you applied it. A second application of the dull spray can provide a matte finish.

PARACHUTE AND STREAMER REPLACEMENT

★ Always use parachutes and streamers that are in good condition.

If you damage a parachute or streamer (which is not unusual), it is simple to make your own. Parachutes can be made from sheets of Mylar,

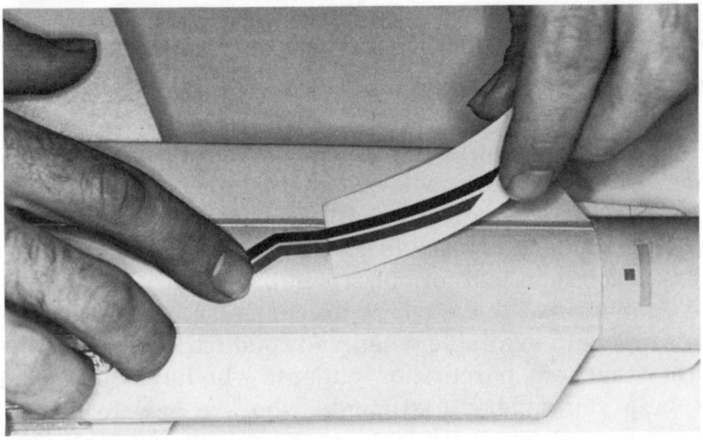

Figure 2-16. Proper application of decals. When all paint is dry, apply the decals in the positions shown. (A) Cut only one decal at a time from sheet. (B) Submerge decal in lukewarm water until decal slides on backing paper (usually 15 to 30 seconds). (C) Gently slide decal from backing paper onto model. (D) Move decal into exact position and carefully blot away excess water with a soft cloth. (E) If the decal "sticks" before you have it in position, apply water over the decal with a brush. This will permit the decal to be moved. (F) Smooth out all wrinkles and air bubbles before the decal dries. We recommend that the completed model be sprayed with Testor's "Dull-Cote." This is a clear flat spray paint that kills the decal shine and protects the model's finish. Photograph by J. Kevin Wolfe, ARTifacts.

from dry cleaner's plastic bags, or from common plastic trash bags. Streamers can be made using flame-resistant crepe paper. Make the parachutes or streamers just as you would make them from the materials in a kit. If you would like stronger shroud lines, you can use nylon fishing line or a heavier cotton string than is supplied wth the kits. Vinyl electrical tape can be used in place of the tape discs supplied with kit parachutes and streamers.

POSSIBLE MODIFICATIONS

★ Never launch an original-design model rocket or a modified kit that has not successfully passed a stability test.

Various shapes of fins and nose cones are supplied with model rocket kits, and these are usually the best shapes for the configuration of the model in question. Some modification is usually possible, however, to increase performance.

Nose cones can be sanded to a more rounded shape, which will sometimes result in decreased drag, because the "ogive" (pointed arch) shape creates the least aerodynamic drag.

SAFETY

★ Plan all activities with safety in mind.

Safety is an important factor, especially when you are working with a group of youngsters. Cutting tools are particularly dangerous in uncoordinated fingers, and it is suggested that you use die-cut balsa or precut the balsa for young students, if possible. Any paper that has to be cut can be cut with scissors, a common tool for classroom use.

If you cannot provide die-cut or precut balsa parts for your students, try to have some help available when you conduct this part of the building process. Other teachers, parents, or students who have built model rockets before can help you supervise students who are new to model building. Supervision is a must when students are using cutting tools.

Before the students begin the assembly, familiarize them with the various components of the model to be built. Hopefully, you will have already spent an hour or two familiarizing your students with rocketry in general, perhaps conducting a demonstration launch in addition to presenting a lecture and/or a film. You should have available a completed model as well as an unbuilt kit. Show the parts of the kit to the students, naming each part and explaining its functions. Present the completed model as a guide. Let each of the students handle and examine the completed rocket to give them a "feel" for what they will be doing. The age and maturity of the group will be your guide in determining how much detailed instruction is necessary.

After you have explained the kit, pass out the students' kits. Tell them to familiarize themselves with the parts in the kits by comparing them with the construction diagram. Then, as you pass out the supplies and tools (glue, sandpaper, and so on), tell the students to look over the assembly instruction sheets. Before proceeding with the actual assembly of the kits, make a dry run through the assembly steps. Begin with the first step, reading aloud the directions and performing the assembly operations in front of the class. Urge students to do the same. This step takes little time and may prevent problems later.

After this, you can begin the actual construction of the rockets. This can be done step by step, or you can let the students proceed at their own pace. (See Figure 2–17.) The step-by-step method is the surest, but you will find that some students work faster than others and will not want to wait while the rest of the class catches up. Allowing students to proceed at their own pace is usually the better of the two methods, because you will have some experienced rocketeers in almost any class. These experienced rocketeers may make excellent helpers. Before you begin assembly, you can ask who is an experienced rocketeer. Enlist these students to help students with problems after they have finished their own kits.

Figure 2-17.
Youngsters building model rockets.
(Photograph by J. Kevin Wolfe, ARTifacts.)

If you have the time, you may wish to illustrate the more difficult assembly steps on the blackboard. This will allow you to show the students, on a group basis, what to do.

You may not wish to help your students this much, especially if you have older youngsters or a gifted class. Reading and following the directions is an important part of the learning involved in model rocketry.

The problems most frequently encountered during assembly involve, as you may already have discovered, gluing operations. Engine mounts are particularly difficult to work with because they have a tendency to "lock up" in the wrong place when they are inserted too slowly into body tubes. With younger groups, you may have to perform this operation for each student. Let the students do their own work whenever possible. This contributes to their sense of accomplishment and their learning. Keep a close eye on all students during any steps requiring gluing so that you can catch and correct errors, such as mislocated fins, before the glue dries.

If a student somehow glues the wrong parts together or puts a fin or engine mount in the wrong place, you can sometimes disassemble the parts by carefully putting water on the glue joints, assuming you are using white glue or some other water-soluble glue. Saturate the glue as thoroughly as possible with water and gently pull on the parts until they separate. Then wipe away the water and excess glue and wait for the parts to dry before reassembling them.

Broken or missing parts are other possible problems. The easiest way to solve this type of problem is to have a couple of extra kits available so that you can immediately replace parts. Or, you may want to stock up on various sizes of body tubes and other parts or invest in a "designer's special" collection of parts.

You can sometimes repair parts, such as broken fins or torn body tubes, with tape or glue. Students will usually not be overly concerned with the appearance of such repairs, but will be happy just to have a completed model.

FINAL CHECKS

★ Use a suitable checklist to make certain that every model rocket has been properly built and prepared for flight before it is launched.

Do not neglect safety checks on launch day!

Once all the kits are assembled, you should perform a safety check on *each* model, concentrating on the following:

- Is the nose cone loose enough for easy deployment of the recovery system?
- Is the shock cord tied securely to the nose cone and firmly fastened to the interior of the body tube by the shock cord mount?
- Is the parachute or streamer properly constructed and attached to the nose cone or shock cord?
- Are the fins, launch lug, and any other external components firmly attached?
- Are all balsa-body tube and balsa-balsa joints filleted with an extra layer of glue?
- Is all the glue dry?
- Are the engine mounts properly installed and firmly in place?
- If the model passes these tests, it is probably ready to fly.

As a final check of the rockets, you may wish to conduct a launch preparation in the classroom. Using your completed model, show the students how to "prep" their rockets.

This will include placing the proper amount of recovery wadding (see Figure 2–18) in the body tube and gently moving it to the bottom of the body tube just above the engine mount.

Be very careful to show the class how to properly prepare the recovery system. Emphasize that the recovery system must be repacked again just before the actual launch.

Demonstrate proper positioning of the nose cone. Mention that care must be taken to make certain that neither shroud lines, shock cord, or anything else can come between the nose cone and the body tube. The fit must be loose enough for the nose cone to easily pop off when the ejection charge functions.

Place an igniter carefully into the engine. Make certain that the igniter touches the propellant at the bottom of the nozzle. Firmly tape the igniter in its proper place.

Finally, demonstrate how to properly secure the engine (with igniter in place) in the engine mount.

Figure 2-18. Blowing recovery wadding into place. (Photograph by Douglas Cannon.)

3
FLYING MODEL ROCKETS

SAFETY CONSIDERATIONS

★ Always follow the Model Rocketry Safety Code. The effectiveness of this set of simple rules has been proven in over 200 million safe model rocket launches.

LAUNCH SITE

★ The launch site must provide a suitable launch area, a convenient preparation area, usable tracking sites, a spectator area, and a recovery area. The flight paths of the model rockets launched must not take them over the spectator or preparation areas.

The launch site selected should be big enough to set aside the preparation and launch areas with enough space left for the recovery area. (See Figure 3–1.) The recovery area should be large enough so that there is no danger of a model rocket drifting or gliding out of the recovery area on descent. The launch and recovery areas should be free of large trees, electrical power lines, dry grass and weeds, and all other potential problems.

The preparation and launch areas should be located in an easy-to-reach location. Precautions should be taken to prevent spectators and persons other than those in the recovery team and tracking teams from wandering into the recovery area. Only those actually preparing their rockets on the launch pads, those waiting to launch their rockets, and range safety and launch personnel should be permitted in the launch area.

Figure 3-1. Typical launch site.

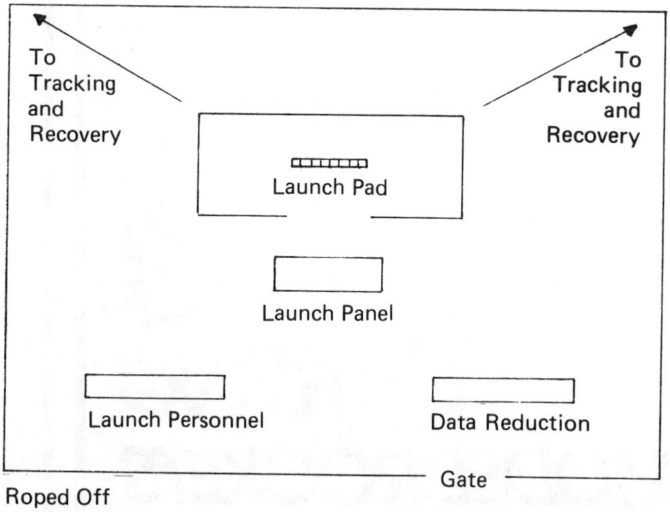

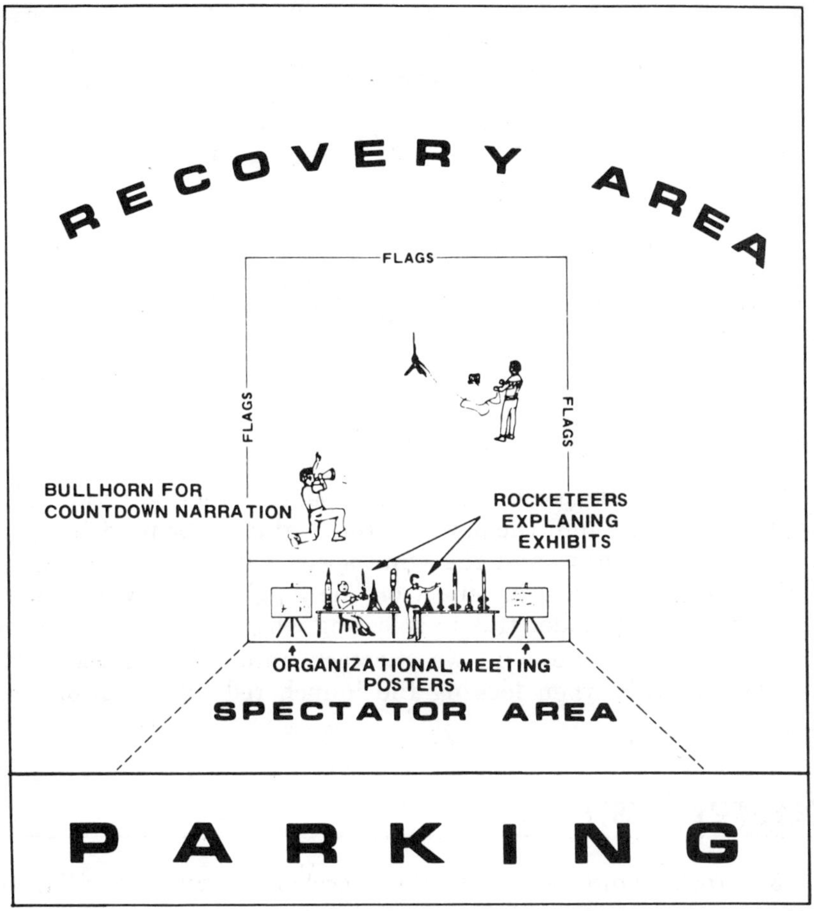

Figure 3-2. Launch site.

The preparation area is where the model rockets have their recovery devices given final preparation, receive final safety checks, and have the engines and igniters inserted. It should be near the launch pads but separate from them. If possible, tables should be provided for the convenience of those prepping their rockets.

The launch pads should be set up so that the rockets will not be over the spectators during any portion of their flight. The launch pads should be positioned in relation to the wind so that there is a minimum chance of the rockets drifting out of the recovery area. The launch personnel, trackers, and spectators should never have to look into the sun while watching the flights of the rockets. See Figure 3-2.

The maximum height to which rockets should be launched will be determined by the size of the recovery area and the speed of the wind. The

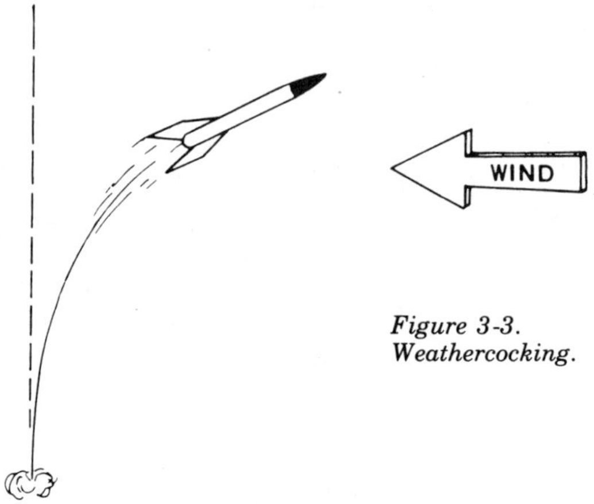

Figure 3-3.
Weathercocking.

higher the rocket is at apogee, the highest point in its flight, the farther it will drift on descent. Limit the power of the engines to be used based on the size of the recovery area and the wind speed. Of course, the weights of different rockets will have an effect on the heights they will reach with specific engines, so consider this in selecting which engines to be used.

A stable model rocket will "weathercock" into the wind (see Figure 3–3), so allow for this when locking the launch rod into position prior to launch.

RECOVERY SYSTEMS

★ Each model rocket launched must have a recovery device that will deploy at or near apogee to bring it safely back to the ground.

Model rockets recovered with parachutes can drift a long way in even a light wind. Gliders can glide great distances if they are trimmed properly. Streamers are very effective recovery devices that permit the safe return of the rockets with minimum drift. Using streamers for small launch areas or in windy weather is advisable.

ENGINE SIZES

★ Each model rocket flight should be made with an engine selected from the manufacturer's list of engines recommended for that model. The engine chosen should be able to cause the ejection charge to activate the recovery system at or near apogee and return the model rocket safely to the ground within the designated recovery area.

The height that a rocket will reach depends greatly upon the power of the engine with which it is flown. (See Figure 3–4.) It is easy to limit the heights that rockets will reach by limiting the power of the engines with which they are flown. Generally, using an engine of the next higher power (for example, by substituting a B engine for an A engine) will cause the height reached by the rocket to nearly double.

In addition to the power of the engine, it is very important to consider the engine's delay time. The ejection charge should eject the recovery device at or just past apogee. If the ejection charge activates the recovery device too soon, the rocket's drag is greatly increased, resulting in the rocket not reaching its maximum height. If the ejection charge operates a few seconds past apogee, the rocket may be falling fast enough that the recovery device will be damaged or detached when it is ejected. If the ejection charge operates more than a few seconds past apogee, the rocket may be falling so fast that the recovery device cannot prevent its crash, even if the recovery device is functioning before impact. That sort of descent doesn't do much for the model rocket on which so much time and effort has been spent, not to mention the cost of the model rocket kit and the model rocket engine.

Figure 3-4. Model rocket engine powers.

Total Impulse Classification

Code	*Pound-Seconds*	*Newton-Seconds*
¼A	0.07–0.14	0.00–0.625
½A	0.14–0.28	0.625–1.25
A	0.28–0.56	1.25–2.50
B	0.56–1.12	2.50–5.00
C	1.12–2.24	5.00–10.00
D	2.24–5.00	10.00–20.00

How High Will Your Model Go?

The chart below shows the approximate altitudes that can be achieved with single stage rockets.

Engine Size	*Altitude Range (depending on rocket size and weight)*	*Approximate Altitude in a typical 1 oz. model*
¼A3-2	50′ to 250′	100′
½A6-2	100′ to 400′	190′
A8-3	200′ to 650′	450′
B6-4	300′ to 1000′	750′
C6-5	350′ to 1500′	1000′

(Some high performance models will reach higher altitudes than shown above.)

The instructions for each model rocket kit contain an exact list of which engines are suitable for launching that rocket. The catalog description of each model rocket kit also lists the engines that are suitable for that model rocket.

LAUNCH PADS

★ The launch pad supports the model rocket in position ready for launch and points it in the direction in which it will be launched.

The launch pad must be large enough and heavy enough to be a stable platform for the rocket. (See Figure 3–5.) The larger the "footprint" of the launch pad, the less likely the launch pad will tilt from a gust of wind striking the rocket, the launcher being placed on an uneven surface, or the rocket being large enough and heavy enough to tip over the launch pad. A heavy launch pad has the advantage of staying securely in place, but it also presents the problem of having to carry around that weight each time the launch pad is moved.

There is a simple, lightweight, and inexpensive means of securing the launch pad very firmly without having to carry a heavy launch pad or place large rocks or other heavy objects on the launch pad legs. This method is to secure the end of each leg of the launcher with hooks made from coat hangers, as shown in Figure 3–6. Two coat hangers yield four of these hooks when the corners are cut off, with about 4 inches of the wire on each side of the corner.

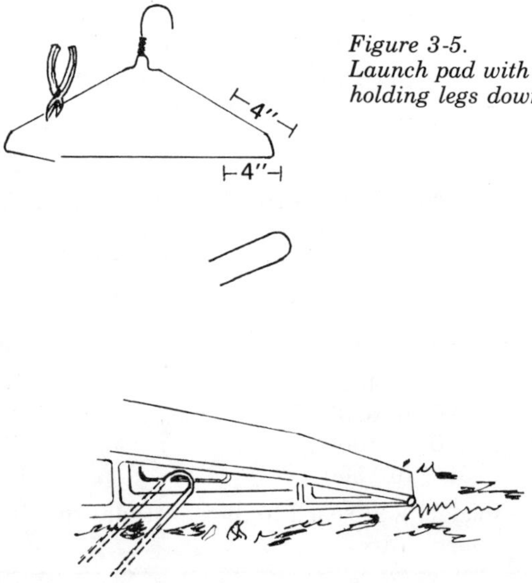

Figure 3-5.
Launch pad with wire hooks holding legs down.

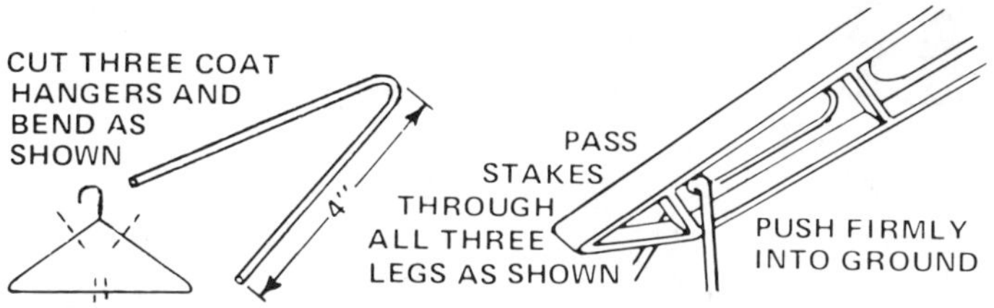

Due to the rocket's length, additional launch pad stability is required for safety. Weigh down or secure your pad with bricks or coat hanger stakes using the following procedure.

Cut three hold-down stakes from coat hangers as shown. Adjust the launch pad on level ground. Pass the stake hooks through the launch pad legs, then push the stakes firmly into the ground. Check to be sure the launch pad is held down securely before proceeding.

Figure 3-6. Hold-down hooks being made from coat hangers.

The launch pad should point the rocket in the direction in which you want it to travel. This is very important because the rocket will leave the launch pad in the direction you point it in. The model rocket is guided in flight by its fins which keep it in the direction it is pointed. When choosing a direction, you should consider the location and size of the recovery area in relation to the launch pad. Another important consideration is the direction and velocity of the wind at the launch site. For launches to very high altitudes, the direction and velocity of winds is very important. You do not want to launch your model rocket over the spectators or over anything on the ground that could present a problem. You do want your rocket to return to the ground within the recovery area. The smaller the recovery area, the more care you must take in adjusting the launcher before the rocket is launched. (See Figure 3–7.)

Each launch pad must have a blast deflector to prevent the exhaust of the rocket engine from striking the launch pad or the ground. Without this blast deflector, the hot exhaust gases of the engine could provide enough heat to cause dry grasses to burn, a wooden base to sear, or a plastic base to scorch.

The launch rod must be firmly secured to the launch pad. The launch rod should be of appropriate thickness, length, and stiffness. The launch rods supplied with commercially manufactured launch pads meet all these criteria. The standard launch rod is 1/8 inch in diameter. An extender section is available for the standard launch rod to use with longer model rockets. Special launch rods or launch rails are available for launching larger model rockets or for special uses. The most common alternative to the standard launch rod is the maxi-rod (1/4 inch in diameter). This maxi-rod is of standard length but is thicker and stiffer. (See Figure 3–8.)

The launch rod should be mounted in a swivel assembly to adjust the angle of the launch rod. This permits the rocket to be launched at an angle of other than 90 degrees. While most rockets are launched vertically, it is

Figure 3-7.
Rocket taking off.

Figure 3-8. Portapad Big foot.

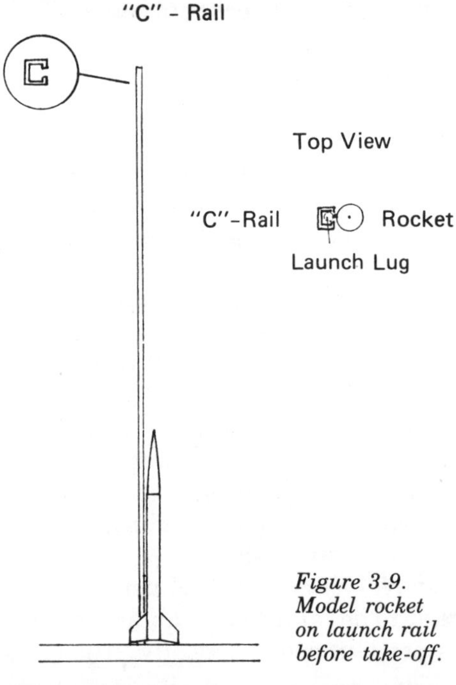

Figure 3-9.
Model rocket
on launch rail
before take-off.

Figure 3-10.
Model rocket
in launch tower.

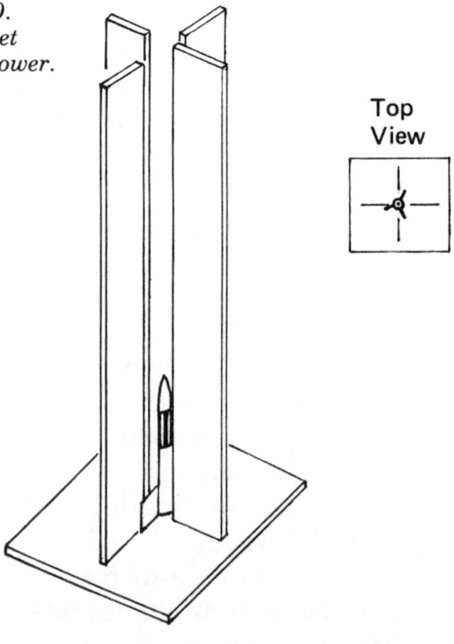

often desirable to launch the rockets slightly into the wind so they can be recovered in the recovery area. The launch rod should never be tilted at an angle of less than 60 degrees (30 degrees from vertical). If the launch rod cannot be adjusted with a swivel, occasionally the entire launch pad can be titled *slightly*, such as by placing a rock under one leg.

A maxirod is a larger, stiffer version of a launch rod. It is 1/4 inch in diameter and is used to launch larger and/or heavier model rockets.

A launch rail is a special device used to secure a model rocket as it is launched. It is suitable for launching large or heavy model rockets. (See Figure 3–9.) The C-shaped section of aluminum is strong enough to be made into very long lengths to provide extra stability at launch. A launch rail can be used to make scale or pseudoscale launch pads.

A launch tower is another method of providing guidance for a model rocket during launch. A launch tower also permits reduction of the drag on the model rocket in flight by allowing the model rocket to be safely launched without a launch lug. The launch tower (see Figure 3–10) requires use of a blast deflector plate.

The piston-tube launcher (see Figure 3–11) also eliminates the launch lug. The rocket is held captive by the inner tube, or "piston," as it rises. The piston is pressurized by the hot gases of the engine's exhaust, and the rocket pops off the piston tube when the tube has traveled its full length.

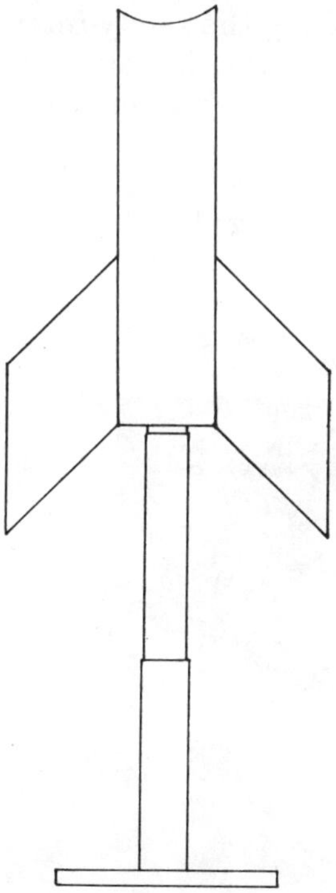

Figure 3-11.
Model rocket on piston launcher.

ELECTRICAL IGNITION SYSTEMS

★ The electrical ignition system provides electricity to the igniter to heat it to start the engine's burn.

The electrical ignition system should be inspected carefully before each launch. Check to be certain that the "jaws" of the microclips are flat, clean, and make good mechanical contact with each other. Placing a small sheet of folded sandpaper (with the abrasive side out) between the jaws and gently moving it back and forth a few times will clean the jaws adequately.

A Solar igniter (see Figure 3–12) requires a minimum of 0.5 amperes of electricity at 6 volts or more. Nichrome igniters require at least 2 amperes at 6 volts or more.

The Solar igniter is composed of two lead wires, a filament wire, a thermo-plastic coating, and a paper reinforcing strip. The special filament becomes very hot and melts when enough electricity passes through it. This filament will reach 550 degrees Fahrenheit as it melts. This heat will cause the special coating on the igniter to burn. This produces extra heat to aid in fast engine ignition. The special coating also helps prevent the leads of the igniter from touching and shorting, which would prevent ignition.

Check the batteries to make sure that each will provide the proper voltage and amperage. The power supply can be given a quick check by clipping the microclips together, then pushing the safety interlock key into position. Do *not* press the launch button. Observe the color of the continuity light. The light should be bright. If the bulb is dim or if it glows orange (instead of white or bright yellow), remove the safety key, open the controller, and test the AA alkaline dry cells with a volt meter to find out which battery (or batteries) is weak and needs to be replaced.

Figure 3-12. Solar launch control system. (Photograph by Michael A. Banks.)

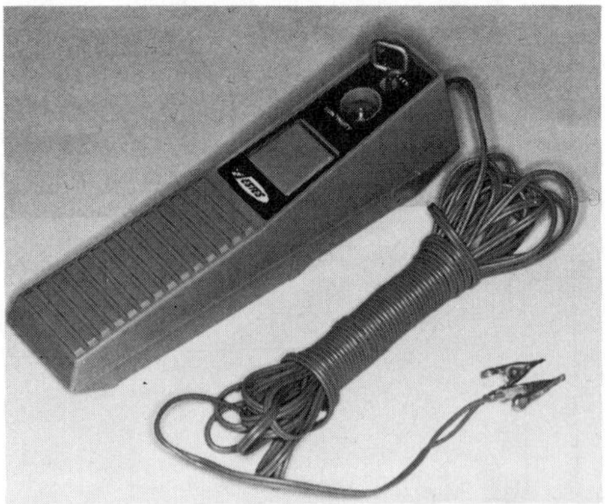

Be careful to avoid using too much electrical power. Do not use household AC current. If you are using a car battery as a power supply, you may wish to use nichrome igniters. Occasionally, the electrical energy provided by a car battery is enough to cause the metal bridge wire of a Solar igniter to heat and vaporize so fast that it does not heat the thermo-plastic coating on the igniter enough to cause it to burn. As a result, the model rocket engine does not ignite.

When using very sensitive igniters, such as the "electric match," the usual launch-control system cannot be used. Even the tiny amount of electrical current that can flow through the continuity light is usually enough to make this igniter function. Be prepared for launch the moment the safety key is inserted should you ever have occasion to use one of these special igniters.

Make sure that the fire button, a momentary contact switch, operates freely and does not stick in the "on" position. Make sure that the metal contacts for the safety key do not make electrical contact unless the safety key is in position in preparation for launch.

Occasionally, an eager rocketeer will be so excited that he or she unconsciously backs away from the launch pad. Since the rocketeer is holding the launch controller, this means that the lead wires are being pulled. If pulled too far, the microclips can be pulled off the igniter leads, or the igniter can be pulled slightly out of contact with the propellant in the engine. If the microclips are detached from the igniter leads, the continuity light will go out, and no ignition can occur. If the igniter is moved, even slightly, it will probably no longer be in contact with the propellant. When the launch button is pushed, the igniter will function, but it will not be touching the propellant, so engine ignition will not occur. This is a misfire. You may or may not be able to detect a small puff of smoke when this occurs. To prepare the rocket for another launch attempt, the safety key must be removed and the ignition system detached. The rocket must then be removed from the launch pad, the old igniter removed, any residue from the igniter in the nozzle removed with a toothpick or similar object, a new igniter inserted and secured, and the rocket again readied for launch.

Having a spare engine of the proper type with an igniter already inserted can save valuable time. When giving a demonstration, this can be an important timesaver. Carefully prepare the engine in the usual manner. Fold the igniter leads over and tape the lead ends to the engine with a piece of masking tape, as shown in Figure 3–13. Make a tab on the loose end of the tape by folding the last half inch of tape back over itself for quickly removing the tape to free the igniter lead ends.

A good way to minimize the chances of the person doing the launching accidentally pulling back and detaching the microclips or moving the igniter is to make a loop in the ignition lead wires near the pad, as shown in Figure 3–14. Insert this loop through one of the holes in a launch pad leg.

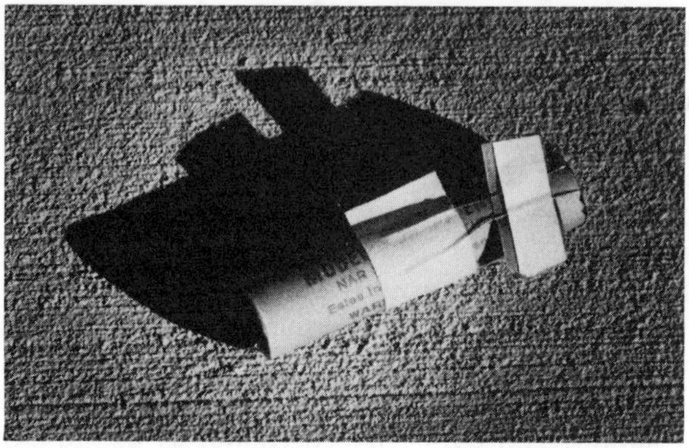

Figure 3-13. Model rocket engine with igniter inserted and igniter leads secured to side of engine with masking tape.

Figure 3-14. Lead wires looped through hole in leg of launch pad.

If the person doing the launching pulls back on the lead wires, you can easily see it, because the loop will be reduced or pulled out. Now you can stop the launch before the wires are pulled enough to move the igniter in the engine.

STORAGE

★ Model rockets and model rocket engines have an indefinite storage period when properly stored.

Figure 3-15. Engines wrapped tightly in a baggie tied closed. (Photograph by J. Kevin Wolfe, ARTifacts.)

Model rockets are lightweight and usually are somewhat delicate. Storing them requires care. Many can be stored in such easily available containers as shoe boxes. Larger or more delicate models can be stored in larger cardboard boxes, such as those available from grocery stores. Suitable boxes are also quite handy for storing kits and models in the process of construction as well as modeling supplies.

One of the best ways to store model rockets is to display them. Your best models will look great on top of a bookcase, a storage cabinet, your desk, a chest of drawers, or similar locations. Model rockets hung by threads attached to straight pins stuck in the ceiling make a spectacular display.

Transporting completed models requires care. If possible, simply carrying the models in your hands is easiest. Placing the models gently on the seat of a car is often adequate if the driver does not turn corners too quickly.

Shipping models requires great care in packing. Using sturdy cardboard boxes with the models carefully packed in Styrofoam™ "popcorn" or similar materials is generally satisfactory. If shredded or crumpled newspapers are used as padding, the models should be placed in clean plastic bags before packing to avoid getting ink on them.

The model rocket engines should be stored to avoid excessive heat, moisture, and jarring. The shelf life of a model rocket engine is many years if the engine is not exposed to excessive heat (temperatures above 130 degrees Fahrenheit or 54 degrees Celsius). Higher temperatures cause the engines to deteriorate. Store the engines in places that will not get too hot. Storing engines in such hot places as the trunk of a car in the sun is not recommended.

In many areas the humidity is no problem. In moist climates with frequent high humidity, carefully wrap the engines in airtight plastic bags (see Figure 3–15) and seal the bags to prevent moisture from the air affecting the engines. Never allow engines to come in contact with water or other liquids. Never launch an engine that is or has been wet. The moisture causes the paper casing of the model rocket engine to swell and weaken. The moisture can also cause the ceramic nozzle to become soft and weak and possibly to crumble away.

Never subject a model rocket engine to strong jarring. While it is very unlikely, it is possible for a strong jarring to cause tiny cracks in the propellant. These tiny cracks can cause uneven burning of the engine with subsequent engine malfunction. This type of malfunction is called a "cato," for catastrophic failure.

PLANNING AND PREPARATION

ASSIGNING RESPONSIBILITIES

★ Specific assignments for each function should be made well in advance of the model rocket launch. Each individual must know his or her responsibilities and be trained to fulfill those responsibilities *before* launch day.

Specific assignments may be made, or individuals may simply cooperate effectively. There must be one person in charge who oversees the entire range for safety.

The operation of the launch is usually divided into different areas of responsibility. The person in overall charge is usually called the *Range Safety Officer.* The actual launch operations are under the direction of the *Launch Control Officer,* usually with the aid of one other person or more. The final preflight checks are carried out by one or more individuals under the direction of the Range Safety Officer. The rockets are recovered by the *Recovery Crew,* unless other arrangements have been made. The height determinations, if made, are carried out by the *Tracking Crews,* usually two teams, each consisting of one or two individuals, and the *Data Reduction Crew,* who calculate the actual heights reached from data supplied by the Tracking Crews. The initial setup of the range and the final policing to remove all trash is done under the direction of the Range Safety Officer, who usually assigns several people to each task. The Range Safety Officer (see Figure 3–16) supervises the entire range and all operations. While his or her primary responsibility is safety, he or she assigns and supervises a number of individuals engaged in a variety of tasks.

The Launch Control Officer and any helpers closely control the area near the launch pads. Models may be placed on the launch pads only when the range is closed to firing and only under the direction of the Launch Control Officers and assistants. Each model must be checked after it is on the assigned launch pad and its igniter connected to the electrical ignition

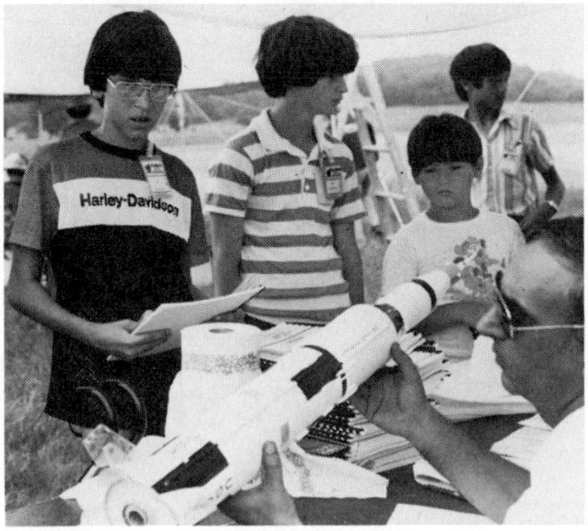

Figure 3-16. Range Safety Officer checking rocket being placed on pad.

Figure 3-17. Launch. (Photograph by Michael A. Banks.)

system. The model must slide smoothly on the launch rod. The microclips must be firmly attached to the igniter leads.

Once each model has been approved for launch, everyone moves back to their assigned positions, and the Launch Control Officer declares the range open for rocket firing. As authorized by the Launch Control Officer, each person launches his or her model rocket following the countdown. (See Figure 3–17.)

Once all models have been launched, the Launch Control Officer announces that the range is closed to further firing before allowing the next rocketeers to approach the launch pads. All safety keys must be removed from the electrical ignition systems before this can be done.

The Recovery Crew is responsible for following the flight of each model rocket and returning it to the launch pad area. No one other than the Tracking Crew, Recovery Crew, Launch Control Officer, and Range Safety Officer should be allowed in the recovery area.

As each model rocket is recovered, it should be returned to the launch pad area so that it can be reclaimed by its owner. The Range Safety Officer should instruct Recovery Crew members *in advance* whether or not they should attempt to recover model rockets that land outside the recovery area. Each rocketeer should have his or her name in or on his or her rocket, such as on a small piece of gummed paper inside the body tube opposite the shock cord mount and *below* the depth that the shoulder of the nose cone reaches.

The Tracking Crew is responsible for making the measurements used to determine the heights reached by the rockets. Normally, each crew needs to determine one number, the exact distance between the tracking station and the launch pad, in advance of the first flight. This number must be accurate. The measurement may be in feet or yards, or in metric units.

Figure 3-18. Altitrak in use.

The most common method of determining this "baseline" distance (distance between the launch pad and the tracking station) is to mark the location of the launch pad and the tracking station in advance of the actual launch day. A small spot of spray paint will show well on grass until the grass is mowed. A steel measuring tape is preferred, but a cloth one may be used if it's in good condition. When several launch pads will be used, a single spot in the center of the launch pads is marked, and all measurements are made with the assumption that this is the launch pad site.

The apogee of each flight is measured with a device that measures the angle between horizontal and the apogee of the flight as observed from the tracking station. This device measures this angular distance in degrees. The exact height reached by the rocket may be determined directly from one of these devices, but the more common procedure is to simply report this angle to the Data Reduction Crew. The Data Reduction Crew uses this angle and the length of the baseline to determine the height of the rocket at apogee, as shown in Figure 3–18.

The device used to determine the angular distance that the rocket reaches may be a surveyor's theodolite. Using this device is difficult because of the narrow angle of vision and the difficulty in moving the device rapidly enough to follow the ascending rocket and then "freezing" the angle when the rocket is at apogee. Also, a theodolite is expensive. An accurate and inexpensive substitute is the AltiTrak™, made by Estes Industries.

For greater accuracy in tracking, it is recommended that two tracking stations be used. Tracking will be covered in more detail in the next chapter.

Runners who carry the written angular measurement(s) for each flight may be used to send tracking data to the Data Reduction Team. Walkie-talkie radios are faster and more convenient. Each number should be read back to the sender to improve accuracy.

The Range Safety Officer will delegate the care of all equipment to specified individuals. These people will be responsible for maintenance of the equipment assigned to them and will make certain all needed equipment is brought to each launch and properly cleaned and stored after each use.

For the usual class or club, it is recommended that each person be trained in performing each responsibility and that the members of the class or club rotate assignments.

Set up the launch pads, preparation tables (if available), and electrical ignition systems so that the control units are grouped reasonably close together between the launch pads and the preparation area. Designate and somehow mark the edges of the preparation area, the launch area, the recovery area, the spectator area, and, if necessary, the parking area (see Figure 3–19). Assign responsible people to make certain that everyone not directly involved in a launch is in the spectator area.

Figure 3-19. Elaborate launch site.

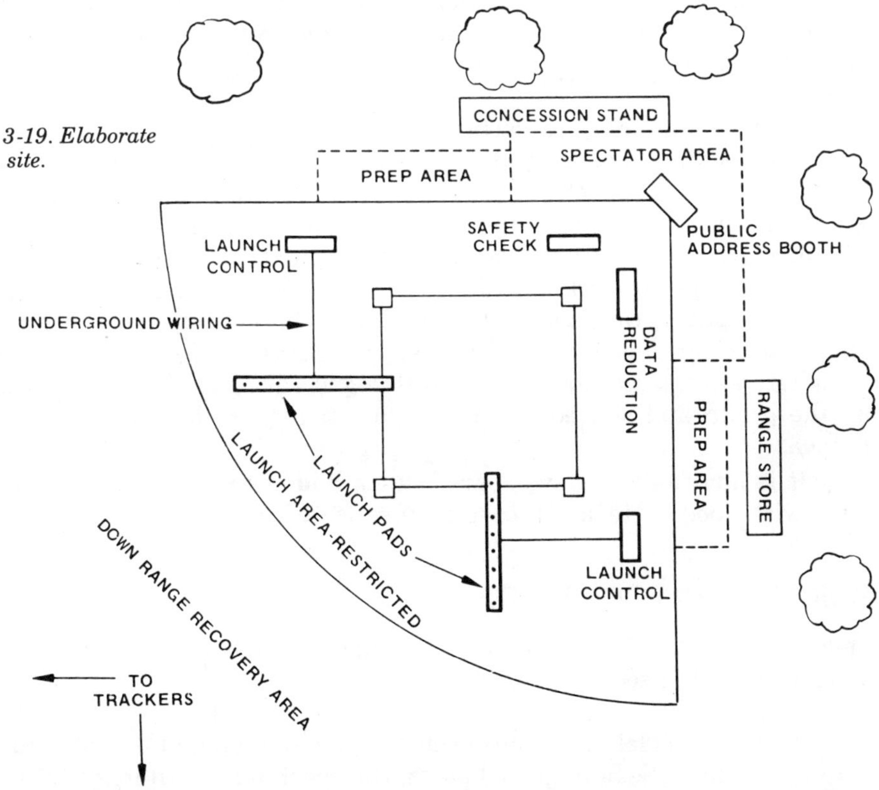

SCHEDULES

★ The exact sequence of activities and the time for each will be clearly posted well before the launch, and everyone concerned will be made aware of this schedule. The schedule will be followed.

Have all parts of the range set up before the launch is to begin. Assign people to take care of this and to make certain that nothing is disturbed. Bring out the group at the appropriate time. Allow time for the people who will be launching first to prep their rockets. Then let this group ready their rockets on the launch pads while the next group prepares their rockets. Everyone not directly involved in the launch usually wants to watch the launches. Let the first group launch. Then let the next group launch while the following group preps their rockets, and so on.

SUPPLIES

★ Each person is responsible for the safety of his or her own model rocket. The care of equipment such as the launch pads, trackers, and similar items shall be as assigned by the Range Safety Officer. Issuing engines, igniters, recovery wadding, and so on, shall be handled as determined by the Range Safety Officer.

Some groups rely on each person to supply his or her own engines, recovery wadding, spare igniters, or repair supplies. Other groups either supply all necessary items or at least keep a small stock of extra engines, igniters, and recovery wadding that individuals may purchase as needed. The most commonly needed extras include model rocket engines appropriate for the types of rockets being launched, Solar igniters, recovery wadding, parachutes, tape discs for parachutes, launch lugs, white glue, and "magic glue" (cyanoacrylate). Basic supplies such as needle-nose pliers, extra safety keys, talcum powder, and fine grit sandpaper are usually owned by the group and shared, with one person responsible for them at each launch.

It is advisable to have spare launch pads and electrical ignition systems with good batteries in case of unexpected need.

PREPARATION FOR LAUNCH

★ Make certain that the model rocket is well constructed with every part in its correct place and securely attached.

After the model rocket has been completed, it should be tested to make sure that it has all the required parts, that each part is in its correct place, and that each part is securely attached. This test should be very carefully performed. Gently attempt to wiggle each fin to make sure that it is securely attached. If a fin wiggles, then it must be securely reattached to the rocket before the rocket can be flown. Make certain that the launch lug or lugs are in the proper place and are firmly attached. If the individual who inspects the rocket for this initial safety test is uncertain of the exact placement of the parts, he or she should check the completed kit against the instructions that came as part of the kit.

In the event that the model rockets are to be launched before they are painted and decaled, make sure that the glue has thoroughly dried before attempting the initial safety check. If the rocket was assembled with white glue, unless the glue has thoroughly dried and turned clear, do not even attempt the initial safety test. The rocket is not ready for flight.

After the rocket has passed the initial safety test, carefully prepare the recovery system as directed in the instructions. Most model rockets use either parachute or streamer recovery. Insert the proper number of sheets of flameproof recovery wadding carefully into the rocket's body tube. Do not pack the sheets of recovery wadding too tightly, or they may act as a cork when the ejection charge is activated. If the recovery wadding is packed too loosely, the hot gases from the ejection charge can leak through the poor seal formed by the loose recovery wadding and scorch or melt the parachute, streamer, or shock cord.

There are several ways to pack a parachute that will provide a maximum opportunity for the parachute to properly eject and open when the ejection charge operates. The instructions on parachute packing that came with the kit are effective. Figures 3–20 show a general technique that is quite effective.

Fully open the parachute. Lightly dust both sides of the parachute with talcum powder. This step is especially important if the rocket is to be launched in cold or humid weather.

Applying a light coating of talcum powder to both sides of the parachute also greatly reduces the static electricity ("static cling") buildup on the parachute. This permits the parachute to open more quickly and easily.

Spread and flatten the parachute until it is folded in half in a semicircular shape. Form the parachute into a spike by folding the parachute in thirds while holding the center by folding in the "sides." Smooth the parachute in its folded condition. Fold the parachute in half by folding the center (top of the spike shape) down over the bottom of the parachute. Fold most of the shroud lines inside the parachute as you carefully roll the parachute into a small cylinder.

Insert this cylinder carefully into the body tube above the recovery wadding. The shroud lines may be *loosely* wrapped around the parachute before inserting it in the body tube, if desired. Gently slide the parachute down into the body tube. Carefully place the shock cord and any remaining

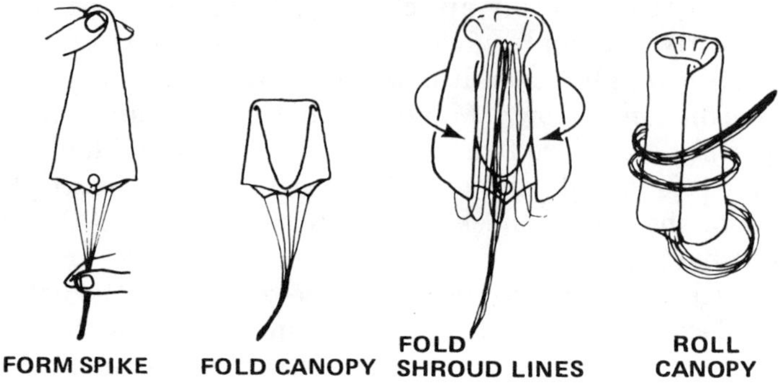

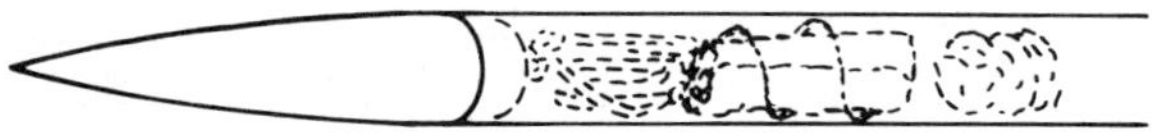

Hold the parachute at its center and pass the other hand down it to form a "spike" shape. Fold this spike in half. Fold shroud lines back along parachute and then back down to lower edge of parachute to reduce length of shroud line "left over". Roll parachute into tube shape to fit easily into body. Any remaining shroud line should be <u>loosely</u> wrapped around parachute. Pack 'chute into the body tube on top of the wadding. Pack the shroud lines and shock cord in on top of the parachute and slip the nose cone into place.

NOTE: DO NOT pack parachute until you are actually ready to launch. For maximum parachute reliability, lightly dust the 'chute with ordinary talcum powder before each flight, especially in cold weather.

Figure 3-20. Parachute preparation.

lengths of the shroud lines over the parachute and gently press down, if necessary, to make room for the nose cone to fit into position without causing pressure on the parachute.

Never leave the parachute packed in the model rocket for more than five minutes, even in warm weather. Never leave the parachute packed in the rocket for more than one minute in cold weather. The plastic of the parachute may "take a set" (become somewhat set in shape) and not properly open when ejected if it has been packed too long. Carefully repack the parachute just before launch.

Reassemble the rocket and make certain that the recovery system can be deployed easily when activated. Select the proper engine based on the engines recommended by the kit manufacturer and the height that the model rocket is to reach on this flight.

Do not remove the paper tape from Solar igniters. Separate the igniters by cutting the paper between them (see Figure 3–21). Leave the paper tape on the igniter to hold the igniter in shape and provide strength. Insert the igniter into the nozzle *until the tip of the igniter is firmly in contact with the propellant at the bottom of the rocket's nozzle.* Hold the igniter firmly in place fully inserted into the nozzle and gently bend the igniter leads over so that the igniter is "folded" over the edge of the engine with the igniter touching the propellant at the bottom of the nozzle. Firmly tape the igniter leads into position with a short piece of masking tape. Push *firmly* to "mold" the tape down around the igniter leads. For insurance, add two additional layers of tape firmly molded down over the igniter leads to hold the igniter securely in place.

An alternative method of securing the igniter is to insert the igniter as described, fold over the igniter leads, and then insert a small amount of recovery wadding rolled into a "pellet" into the nozzle and pushing it *firmly* down to act as a cork holding the igniter securely in place.

Inserting the electrical igniter into the engine is perhaps the most critical step in the prelaunch procedure. About 90 percent of all misfires (the igniter burns, but the engine does not ignite) are caused by the igniter not being in its proper place in the engine at the moment the electrical current for ignition passes through the igniter.

After carefully securing the igniter in the engine, gently but firmly insert the engine into the rocket body tube as specified in the instructions. The engine must remain firmly in place during thrust, coasting, ejection, and recovery. Some engines are held in place by an engine hook. Sometimes engines are secured by making a tight friction-fit between the engine and the body tube by applying masking tape around the engine before it's inserted in the body tube. Occasionally, the instructions in a kit will call for the engine to be taped to the rear of the body tube by wrapping masking tape around both the rear edge of the engine and the rear edge of the body tube.

The final safety check before the rocket is placed on the launch pad makes certain that the recovery system is properly packed; that the nose cone is tight enough not to wobble yet loose enough not to stick when the ejection charge functions; that the proper engine has been selected, properly

Figure 3-21. Solar igniters being cut apart.

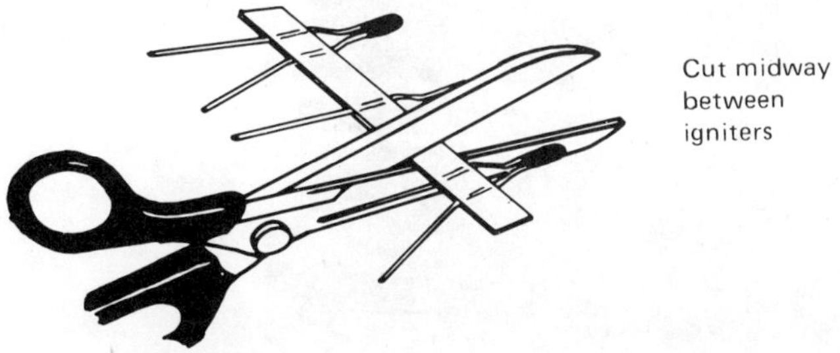

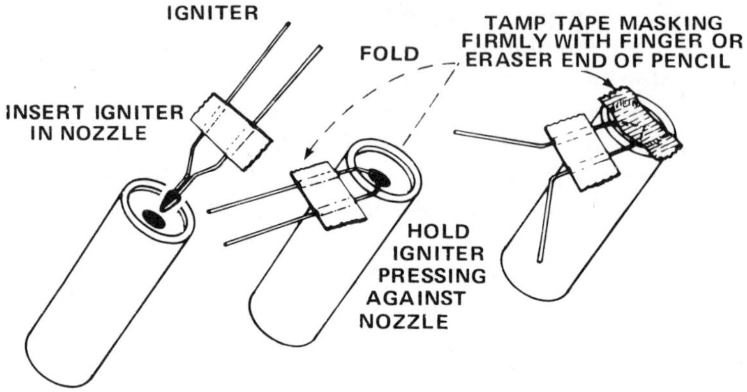

Figure 3-22. Solar igniter properly inserted and secured in engine.

inserted, and retained; that the igniter is securely in place; and that the model's center of gravity with the engine in place is in its proper position.

Do *not* allow any model to be launched that is not properly constructed or properly prepped for flight. The individuals performing the safety check should be carefully trained and should be very strict in checking each rocket carefully.

In addition to checking the construction of each model and the preparation of the recovery system, make certain that a suitable engine has been selected and that the igniter is in place and well secured (see Figure 3–22).

Figure 3-23. Swing test.

SWING TEST

★ The swing test, or string test, is a simple way to predict if a model rocket will be stable in flight. Prepare the recovery system in the usual way and insert the model rocket engine into the model rocket. Attach a long (4 to 6 feet) string with a slip loop at one end to the model rocket. (See Figure 3–23). Balance the rocket on the string, then use a short strip of masking tape to secure the rocket on the string at the rocket's center of gravity. Then swing the rocket around your head in a large, open area, as shown in Figure 3–23. If the rocket's nose remains pointed ahead as you swing the rocket around your head, the rocket is probably stable. Always use the heaviest engine that will be used in flying for this test. The heavier the engine, the further to the rear is the center of gravity.

LAUNCH RANGE OPERATIONS

EQUIPMENT SETUP

★ Well before the launch is to start, the launch area, the preparation area, the tracking stations, the spectator area, and any other areas should be clearly marked.

All launch pads, electrical ignition systems, tracking equipment, and other necessary items should be set up and rechecked. Make any necessary adjustments, repairs, or replacements. All assigned personnel should be at their stations well before the launch is scheduled to begin.

Set up the launch area so that the model rockets will not fly over the spectators, even if the rockets weathercock into the wind. Be certain that neither the trackers, the launch crew, or the spectators will look into the sun while observing rocket flights.

Preparation of launch equipment has been covered earlier in this chapter.

LAUNCH ANGLE AND DIRECTION

★ The launch rods should be pointed either vertically or slightly into the wind. The model rockets should not be launched over the preparation or spectator areas. The launch rods should never be angled more than 30 degrees from vertical.

If the launch angle is less than 60 degrees from horizontal, a rocket with too long a delay element could strike the ground before the ejection charge deploys the recovery device.

The Launch Control Officer and assistants should check the direction and angle of the launch rod on each pad before each launch to make sure that the model rocket will not be launched over the spectator or preparation areas and not drift outside the recovery area. Always be aware of the direction from which the wind is blowing and the approximate velocity of the wind. These factors must be considered for *each* launch. Do not launch in high winds (winds blowing over 20 miles per hour).

PREFLIGHT CHECK

★ Each model rocket to be launched should be checked for proper construction before it is prepped.

Check the engine chosen for the launch to be certain that the engine is one of those specified for flying that model rocket and that the engine is appropriate for the conditions prevailing at the time of launch. Check to be certain that the igniter is properly inserted and secured. Be certain that the engine is properly inserted and secured, that the recovery system is properly prepped, and that the center of gravity of the model is in an acceptable location before the rocket is placed on the launch pad.

COUNTDOWN AND LAUNCH

★ Make certain that the model rocket's launch lug slides smoothly on the launch rod. Be sure that the micro-clips are securely attached to the igniter leads. Launch within five minutes (one minute in extremely cold weather) of prepping the recovery system. Clear the launch area before start of the countdown. Launch the model rocket when the countdown reaches zero.

The Launch Control Officer should make certain that the microclips are carefully cleaned at least every five launches. The Launch Control Officer or whoever is designated to launch each model should give the countdown *loud and clear* from five to zero. At "zero," the launch button should be depressed and held down until the model rocket engine ignites. If ignition does not occur within five seconds, the batteries are dead, a misfire has occurred, or a malfunction has happened. Release the launch button and remove the safety key. Wait 30 seconds, then approach the launch pad, remove the model rocket from the launch pad, and determine the cause of the problem and correct it.

If the cause of the failure of ignition was a misfire, remove the spent igniter and carefully replace it with a new electrical igniter. Make certain that the igniter is inserted all the way to the bottom of the nozzle, then carefully secure the igniter firmly in with masking tape.

If the reason ignition did not occur was a weak battery, replace the ignition system with a spare system for the duration of the launch. If the problem cannot be easily diagnosed, replace the electrical ignition system with a spare system for the duration of the launch.

As soon as the model rocket has been launched, the person who launched the rocket must remove the safety key from the ignition system. The safety key should be turned over to the proper person as soon as it is removed from the ignition system.

TRACKING

★ It is extremely important that the measurement of the baseline, the angle reached by the rocket at apogee, and the determination of the height reached be accurately performed.

Tracking is treated in detail in the next chapter.

RECOVERY

★ Only authorized persons are to enter the recovery area or to recover model rockets after launch.

Members of the Recovery Team must be observant, alert, fast-moving, and careful. Observing a model rocket near apogee can be very difficult. Visually following the rocket on its descent and then going to the rocket on the ground and retrieving it requires alertness and care. It is recommended that the persons involved in the recovery not attempt to catch the model rocket before it touches the ground. There are few sounds more depressing to a model rocketeer than hearing the soft "crunch" of a model rocket being stepped on or otherwise destroyed! Another problem that must be avoided is eager would-be recoverers colliding with each other and/or the rocket as they attempt to retrieve it.

Finding lost model rockets can be very difficult. It is very important that the model rocket be observed carefully during its entire flight. One glance away may be enough to make it extremely difficult or impossible to find the model again in flight. Finding a lost rocket once it is on the ground can be extremely difficult, even on apparently flat, open ground.

Perhaps the best way to find a rocket once it is on the ground is to stop moving before the rocket touches down and then carefully mark the exact direction toward the model at touchdown. Then note an easily recognized landmark near the line of sight to the point at which the rocket touched down. Carefully walk a straight line toward the apparent point of touchdown. The closer the landmark is to the line of sight, the easier this job becomes. As the apparent point of touchdown is neared, slow down and look carefully all around. Unless the downed rocket is positively identified, do not depart from this straight-line path. If the rocket is still not found after the point at which the rocket appeared to land has been passed, the problem of locating the rocket becomes much harder.

Perhaps the best way to proceed is to return to the apparent point of touchdown and reexamine the entire area carefully. If the rocket has still not been found, follow a slowly enlarging spiral as you walk about the area, searching carefully. If the rocket has not been found after the searched area is very large, good luck! The chance of finding the rocket is now very small. This is bad news to the rocketeer who built the rocket, and it certainly won't do much for the popularity of the Recovery Team member.

After recovery, the rocket is returned to the launch site to be reclaimed by its owner. If the rocketeer has another opportunity to launch the rocket, the rocket is carefully checked for damage. If damage is found, the rocket is repaired before being reprepped for launch. Do not launch a rocket on which the glue used in repairs is not yet dry. White glue will turn clear when it is dry. Always check the repair by physically trying to wiggle the parts to make certain that the repair job is adequate. Once the rocket is proved to be in satisfactory condition, the rocket may be taken to the preparation area and reprepped for flight by following the same procedures as used before its first flight.

SUPPORT ACTIVITIES

DOCUMENTING FLIGHTS

★ Keep accurate records on all aspects of each flight that are important enough to document.

The number and types of records kept will be determined by the purpose of the launch. For contests, extensive records are necessary. For casual, fun launches, few, if any, records are necessary. Develop and use an appropriate flight data sheet, as shown in Figure 3–24.

Narrative reports on launches can be newsworthy if carefully written. Often school or town newspapers will carry a report on a model rocket launch if it is well written and includes the names of a number of local residents along with information about where, when, and why the model rocket launch was held. This can provide the opportunity for budding reporters to get their articles into print as well as provide good, free publicity for the school and your model rocketry program.

Photographs of model rocket activities, construction and preparation as well as launches, can be good documentation of what was done for use in

ESTES®

A SUBSIDIARY OF DAMON

Model Rocket FLIGHT DATA SHEET

PUBLISHED AS A SERVICE TO ITS CUSTOMERS BY ESTES INDUSTRIES, PENROSE, COLORADO 81240

© ESTES INDUSTRIES 1968

NAME ____________

ADDRESS ____________

CITY ______ STATE ______

ZIP ______

USE ONE FLIGHT DATA SHEET FOR EACH MODEL. FOUR FLIGHTS MAY BE RECORDED ON EACH SHEET

PREFLIGHT INFORMATION

Rocket Name ______	Nose Cone Type ______
Rocket Number ______	Fin Type ______
Date Completed ______	No. of Fins ______
Type of Rocket ______	Color Scheme ______
No. of Stages ______	Weight Empty ______

COUNTDOWN CHECKLIST

(Checkboxes: Flight #1, Flight #2, Flight #3, Flight #4)

Parachute or Streamer

☐☐☐☐ 12) Pack flameproof recovery wadding into the body tube. Insert the parachute or streamer.

☐☐☐☐ 11) Install the nose cone or payload section. Check condition of the payload (if any).

☐☐☐☐ 10) Apply masking tape to the engine(s) to achieve a tight friction fit in the body tube(s). When launching a multi-stage rocket be sure that the engines are in their proper relative positions and that a layer of cellophane tape is wrapped tightly around each engine joint.

☐☐☐☐ 9) Install an igniter in the engine.

☐☐☐☐ 8) Place the rocket on the launcher. Clean and attach the micro-clips.

☐☐☐☐ 7) Clear the area, check for low flying aircraft, alert recovery crew and trackers.

☐☐☐☐ 6) Arm the launch panel.

☐☐☐☐ 5)

☐☐☐☐ 4)

☐☐☐☐ 3)

☐☐☐☐ 2)

☐☐☐☐ 1)

☐☐☐☐ LAUNCH!

Feather-weight or Tumble

☐☐☐☐ 10) Fit the engine in the body tube carefully so that the recovery system will function properly.

☐☐☐☐ 9) Install an igniter in the engine.

☐☐☐☐ 8) Place the rocket on the launcher. Clean and attach the micro-clips.

☐☐☐☐ 7) Clear the area, check for low flying aircraft, alert recovery crew and trackers.

☐☐☐☐ 6) Arm the launch panel.

☐☐☐☐ 5)

☐☐☐☐ 4)

☐☐☐☐ 3)

☐☐☐☐ 2)

☐☐☐☐ 1)

☐☐☐☐ LAUNCH!

Boost-Glide

☐☐☐☐ 12) Check payload section (if used) and slide it into position.

☐☐☐☐ 11) Set trim adjustments for desired flight path and for existing weather conditions.

☐☐☐☐ 10) Check engine for proper fit in body tube.

☐☐☐☐ 9) Install an igniter in the engine.

☐☐☐☐ 8) Place the rocket on the launcher. Clean and attach the micro-clips.

☐☐☐☐ 7) Clear the area, check for low flying aircraft, alert recovery crew and trackers.

☐☐☐☐ 6) Arm the launch panel.

☐☐☐☐ 5)

☐☐☐☐ 4)

☐☐☐☐ 3)

☐☐☐☐ 2)

☐☐☐☐ 1)

☐☐☐☐ LAUNCH!

LAUNCH

Flight Number		Flight #1	Flight #2	Flight #3	Flight #4
Date of Launch					
Launch Location					
Payload	Description				
	Weight				
Recovery System	Type				
	Color				
Engines No. of / Type	1st Stage	/	/	/	/
	2nd Stage	/	/	/	/
	3rd Stage	/	/	/	/
Total Weight					
Method of Launch					
Launch Angle					
Predicted Altitude					

WEATHER

Wind Direction				
Wind Velocity				
Humidity				
Temperature				
Visibility				
Remarks				

FLIGHT DATA

Altitude	Estimated				
	Tracking Information				
	Computed Alt.				
Flight Duration					
Stability Information					
Flight Performance					

Figure 3-24. Flight data sheet.

Figure 3-25.

bulletin boards, instruction for the next group of rocketeers, and possible publicity for the program in local newspapers, magazines, and school newspapers (see Figure 3–25). Covering activities with a movie camera or videotape camera produces excellent documentation. These media can also be used by local TV stations for additional publicity. Model rocketry activities are very photogenic and are popular as "human interest news."

4
ALTITUDE TRACKING

ALTITUDE DETERMINATION

★ Accurate determination of heights reached requires care in all aspects of measuring, recording, and calculating.

The responsibilities of the Tracking Crew in determining the tracking baseline and angular measurements were covered in the previous chapter.

The baseline distance can be measured using a standard measuring tape. The angular distance moved by the rocket can be measured with any accurate angle-measuring device. The most popular device for measuring heights is the Estes AltiTrak™, shown in Figure 4–1. This device can measure the angular distance risen by the rocket. If the baseline distance is 150 meters, the actual height reached by the rocket can be determined directly from the AltiTrak. If the angular distance is measured, the actual height reached can be determined through use of the angle measured.

The device used for measuring the angular distance reached can be homemade, as shown in Figure 4–2. Such devices are reasonably accurate, but they may not be durable enough for repeated use.

TRACKING TECHNIQUES

★ Either one- or two-station tracking can be used. Accuracy in making and recording all measurements is very important.

One-station tracking is the easiest tracking system to use. The results produced are generally reliable enough for most uses. This system consists essentially of accurately measuring the distance between the launch pad and the tracking station, measuring carefully the angular distance risen by the rocket, and then determining the actual height reached by the rocket through a simple calculation (see Figure 4–3).

One effective way to increase accuracy of tracking is to have two people on each tracking team. The two help each other and check each other's work. Having two tracking teams and taking the average of the two heights calculated is a good way to increase accuracy. If the two results are more than 10 percent apart, both heights are thrown out. This not only provides greater reliability but uses peer pressure to encourage taking careful measurements.

The formula for determining the height reached by a model rocket flight is:

Height = Baseline × Tangent of angular distance

See Figure 4–4. If the tracking station is positioned at A, side b represents the baseline. Side a, the opposite side, represents the height of the

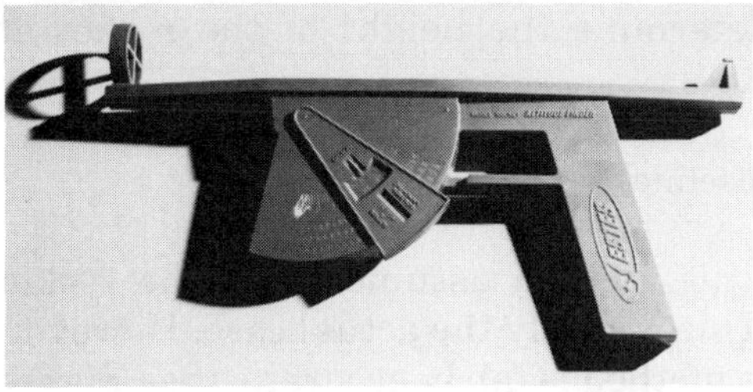

Figure 4-1. Altitrack. (Photograph by J. Kevin Wolfe, ARTifacts.)

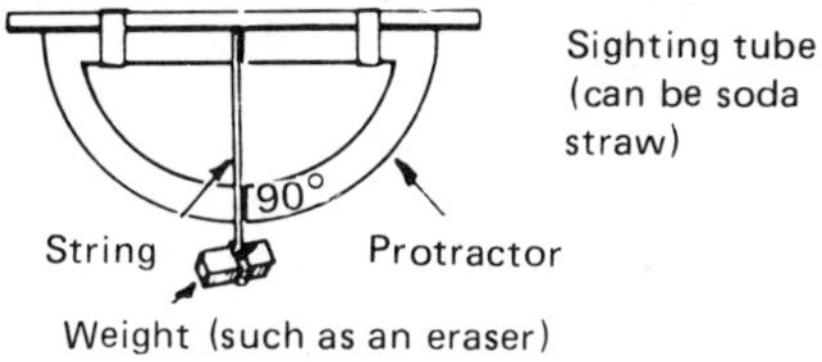

Figure 4-2 (left). Homemade altitude measuring device.

When using the homemade altitude measuring device, angular distance is found by subtracting the reading taken (angle marked) of the rocket at apogee from 90°.

90° - 60° = 30°
angular distance

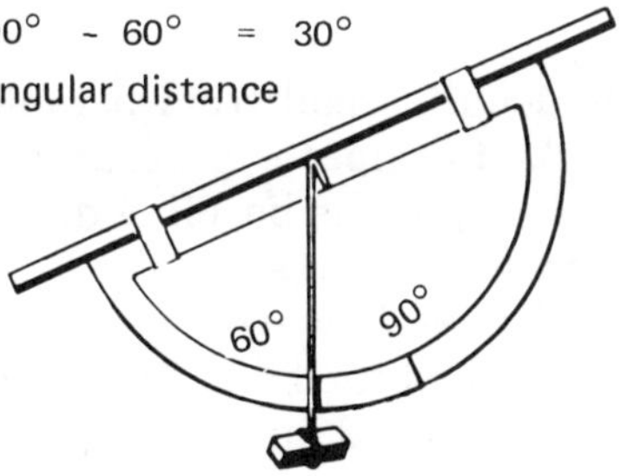

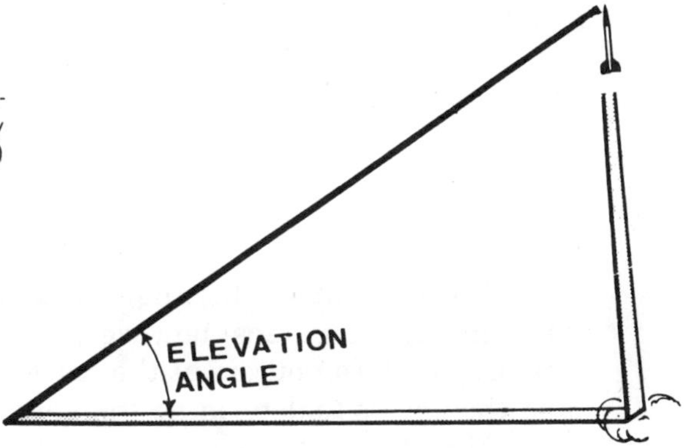

Figure 4-3. Single station tracking.

Figure 4-4. Right triangle with angles and sides labeled.

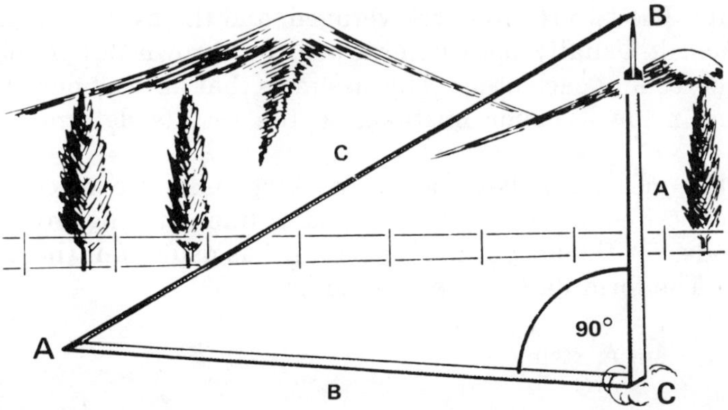

rocket at apogee. The formula to determine the height of the rocket at apogee is:

Height = Tangent of angle A × Baseline

Altitude determination using this method assumes that the rocket rises vertically over the launch pad. This is rarely the actual case. However, if the baseline is reasonably long, this method is fairly accurate. (See Figure 4–5.) Position the tracking station at a right angle to the wind movement in relation to the launch pad to improve accuracy.

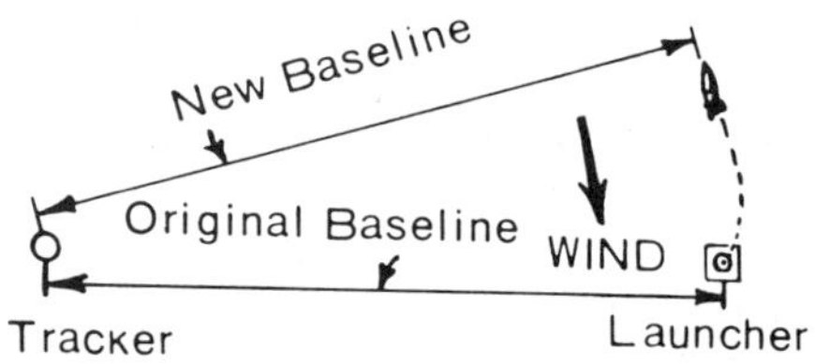

Figure 4-5. Vertical view of tracking station at right angle to wind.

The tangent of the angle measured is usually read from a table of tangents. Some groups, however, have developed computer programs that are used with computers at the range to simply input the baseline distance and the angles to determine the heights reached.

Two-station tracking is more accurate. It does, however, require two tracking stations instead of one. The determination of the height reached by each flight involves determining more numbers and doing more calculations. The simplest method of two-station tracking involves placing the two tracking stations on opposite sides of the launch pad at a right angle to the wind direction (see Figure 4–6). Each tracking station takes its own measurement of the angle reached, and the Data Reduction Team calculates both heights. The two results are averaged, and the average is used as the height reached. Usually both heights will be thrown out if they are not within 10 percent of each other. This assumes that at least one of the tracking stations made a major mistake, so the results determined are not reliable.

Another method of two-station tracking uses two tracking stations placed on opposite sides of the launch pad in line with the wind. This time the two angles for the apogee are determined as usual, but the data is used differently. The formula for this method is:

$$CD = \frac{c \times \sin A \times \sin B}{\sin C}$$

CD Height

c Baseline AB between tracking stations

A Angular distance measured from tracking station A

B Angular distance measured from tracking station B

$C = 180° - \angle A - \angle B$

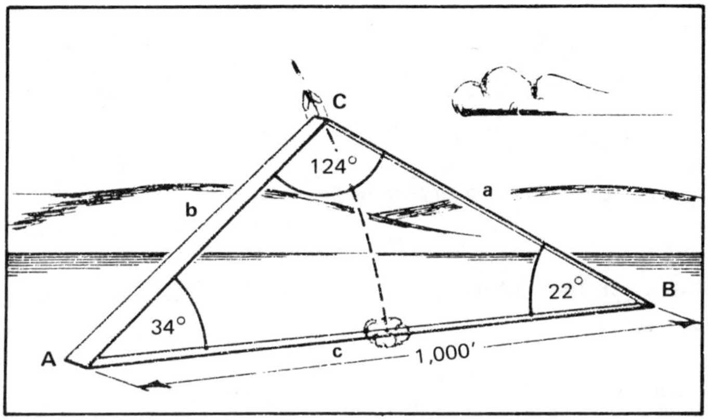

Figure 4-6. Two station tracking without azimuth.

A more complicated but more reliable tracking system that compensates for the rocket rarely being vertical over the launch pad at apogee is the altitude-azimuth system (see Figure 4–7). This system requires making five measurements before the height calculations (data reduction) are made. These five numbers are the baseline length, the angular distance the rocket has risen as determined from one tracking station, the azimuth reading of the rocket at apogee from this station, the angular distance the rocket has risen as determined from the other tracking station, and the azimuth reading of the rocket at apogee as measured from the other tracking station (see Figure 4–8).

Figure 4-7. Azimuth angle. A direction expressed as a horizontal angle from a given reference point.

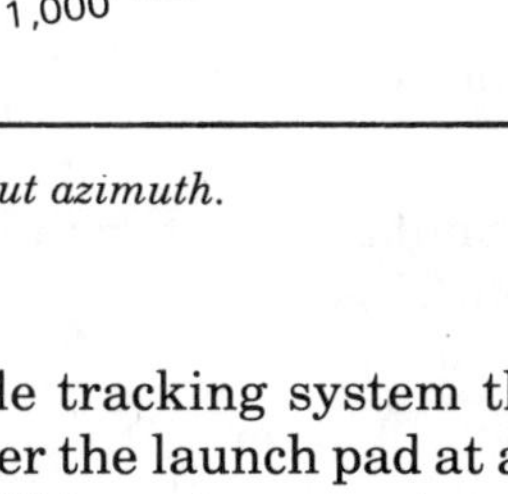

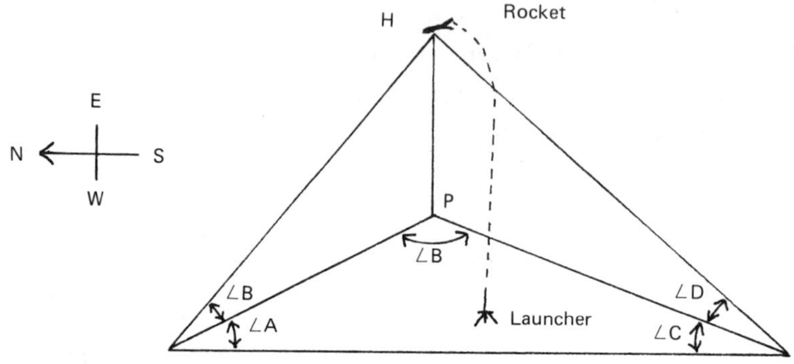

Figure 4-8. Two-station altitude-azimuth tracking.

The formula for determining the height reached as determined by "tracking north" is:

$$PH = \sin \angle C \times \tan \angle B \times \frac{(b)}{\sin\,(180 - [\angle A + \angle C])}$$

The formula for determining the height reached as determined by "tracking south" is:

$$PH = \sin \angle A \times \tan \angle D \times \frac{(b)}{\sin\,(180 - [\angle A + \angle C])}$$

The two heights as determined by tracking north and tracking south are averaged. Then the average is recorded as the height of that flight if neither height calculated is over 10 percent away from the average height. If both heights calculated are not within this range (average height plus or minus 10 percent), the heights are considered unreliable and are not used. The height is recorded as "track lost." Using this system places peer pressure on the tracking teams, because no one wants their height lost because someone on a tracking team or the data reduction team was careless.

Sources of error in determining the heights reached by model rockets include inaccurate angle-measuring devices, inaccurate baseline measurement, inaccurate measurements for angular distance risen and/or azimuth, inaccurate math tables for determining functions of the angles used in determining heights, carelessness in transmitting or copying data, and carelessness in making calculations. It's wise to practice height determination with fixed objects of known height before the actual launch. Practice height determination techniques by measuring the heights of fixed objects such as a certain corner of a building, the knob on the top of a flagpole, and so on, whose heights are known or can be accurately measured by direct means. Once accuracy is reached on these fixed objects, let tracking teams practice by simultaneously measuring the height reached by a baseball thrown vertically into the air. Once the height reached by the baseball can be reliably determined, practice tracking a few times with an actual model rocket launch. After reliability is reached, the tracking teams are ready to perform their function at a regular model rocket launch.

DATA REDUCTION

★ All calculations should be made with great care and all data recorded accurately.

The height reached at apogee by each rocket is determined from measurements made on the ground. For most tracking situations, only three numbers are needed to determine the height reached by the rocket. The baseline, or distance between the launch pad and the tracking station, is determined when the tracking station is initially placed in position. It is convenient to make this distance a round number, such as 200 feet, 150 meters, and so on.

The tracking stations should always be at the same level as the launch pad. While it is possible to compensate for differences in height between these locations, it is inconvenient.

Always place tracking stations (and spectators) so that they do not face directly into the sun or close to it when observing the rocket's flight.

Calculate the height reached by a rocket if one-station tracking provided this set of data. The tracking station is 300 feet from the launch pad. The angular distance risen by the rocket was 50°.

Height = Baseline × Tangent of angle

Height = 300 feet × 1.19

Height = 357 feet

Calculate the height reached by a rocket if two-station tracking provided this set of data. The tracking stations are 200 meters apart. The readings from tracking north were: altitude 50°, azimuth 30°. The readings from tracking south were: altitude 43°, azimuth 20°.

Tracking North

$$\begin{aligned}\text{Height} &= \sin 20^\circ \times \tan 50^\circ \times 200.\text{ m}/\sin(180 - [30^\circ + 20^\circ])\\ &= 0.342 \times 1.192 \times 200.\text{ m}/\sin 130^\circ\\ &= 0.342 \times 1.192 \times 200.\text{ m}/0.776\\ &= 0.342 \times 1.192 \times 261.1\text{ m}\\ &= 106.4\text{ m}\end{aligned}$$

Tracking South

$$\begin{aligned}\text{Height} &= \sin 30^\circ \times \tan 43^\circ \times 200.\ \text{m}/\sin(180 - [30^\circ + 20^\circ]) \\ &= 0.500 \times 0.933 \times 200.\ \text{m}/\sin 130^\circ \\ &= 0.500 \times 0.933 \times 200.\ \text{m}/0.776 \\ &= 0.500 \times 0.933 \times 261.1\ \text{m} \\ &= 121.7\ \text{m}\end{aligned}$$

Averaging the two answers results in 106.4 m + 121.7 m = 228.1 m/2 = 114.1 m.

Since results within 10 percent of the average will be allowed, 114.1 m × 0.9 (10 percent below average) = 102.7 m is below the calculated height of 106.4 m, so this reading is OK. For the other height, 114.1 m × 1.1 (10 percent above average = 125.5 m is above the calculated height of 121.7 m, so this reading is OK. The average height of 114.1 m is recorded as the official height.

To illustrate this another way, see Figure 4–9.

FLIGHT ANALYSIS

Mathematical analysis of expected flight performance, performed either by computer simulations or by hand, can provide guidance to the performance

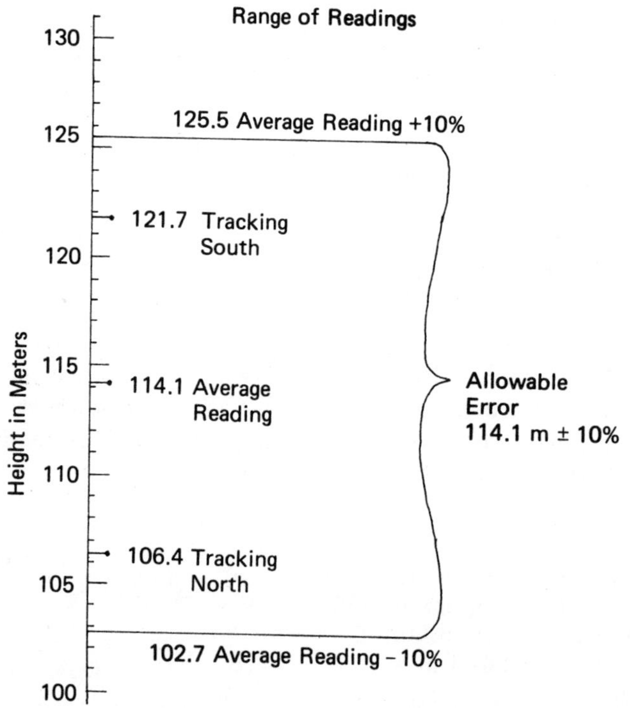

Figure 4-9. Allowable error.

expected by modifications of the rocket's weight, drag, or engine. However, experiments should be performed to validate the theoretical calculations. This involves controlled experiments.

Controlled experiments are experiments that are repeated with only *one* factor changed. Such an experiment might determine the increase in performance when a certain model rocket is launched with an A8-5 engine instead of A8-3 engines. The rocket should be launched at least three times with A8-3 engines and each flight accurately tracked. The same rocket should be launched at least three more times with A8-5 engines and these flights also accurately tracked. Comparing the average height reached by the rocket when flown with an A8-3 engine and an A8-5 engine will then be possible.

The factors that most affect the drag on the rocket are the frontal area of the rocket, the "wetted surface" (total exterior surface of the rocket subject to drag in the air stream), the shape of the rocket, and surface irregularities, such as abrupt changes in body diameter, the surface finish on the rocket, and similar characteristics of the rocket. Fortunately, the drag of the rocket may be markedly altered by the way the rocket is built.

The greater the weight of the rocket, the less altitude it will achieve with a specified type of model rocket engine. The greater the drag on the rocket, the less altitude it will achieve with a specified type of model rocket engine.

Variables in flying conditions that can affect flight performance include air temperature (the lower the temperature, the denser the air), wind (wind produces a lateral force on the rocket that tends to make the rocket weathercock, producing lateral motion and resulting in less height), nearby thunderstorms (don't fly!), and extremely cold conditions (no fun to fly unless you are really bundled up. Also, recovery systems are more likely to fail).

5

BASIC PHYSICS THROUGH MODEL ROCKETRY

FORCES ON FLYING OBJECTS

★ The four basic forces operating on objects in flight are gravity, thrust, drag, and lift.

All objects in flight are subject to these four basic forces. (See Figure 5–1.) The actual movement of the object is determined by how the forces work together.

Gravity pulls all objects toward the center of the earth. The amount of this force is proportional to the mass of the object. The distance between the object and the center of the earth is also important, but this can be neglected for objects such as model rockets, which are operated near the surface of the earth.

Thrust is the force that propels the flying object. (See Figure 5–2.) For a model rocket whose engine is producing thrust, this thrust is the force pushing on the flying object. The thrust is directed away from the direction of flight. The other forces continue acting after the thrust has ceased.

Drag is the force acting on an object moving through a fluid. Air and water are fluids. Drag is the resistance that the object encounters as it moves through the fluid. Drag is discussed in more detail in Chapter 6.

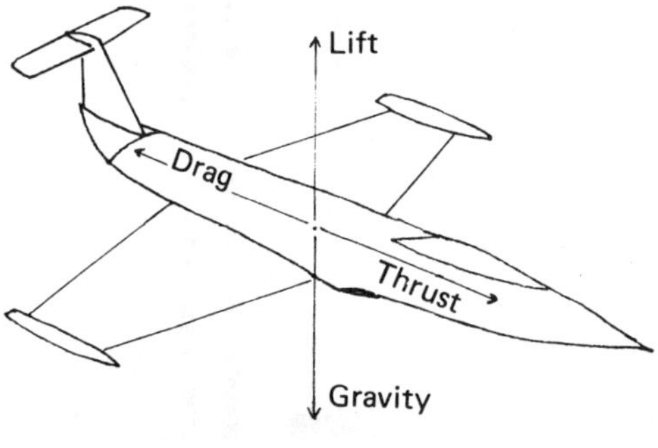

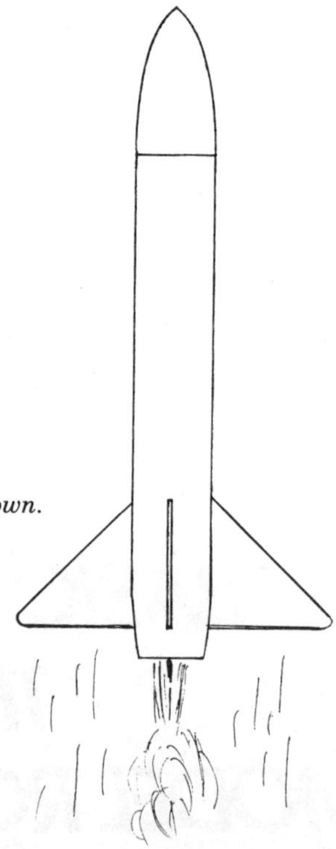

Figure 5-1 (above). Object in flight with four forces shown.
Figure 5-2 (right). Model rocket under thrust.

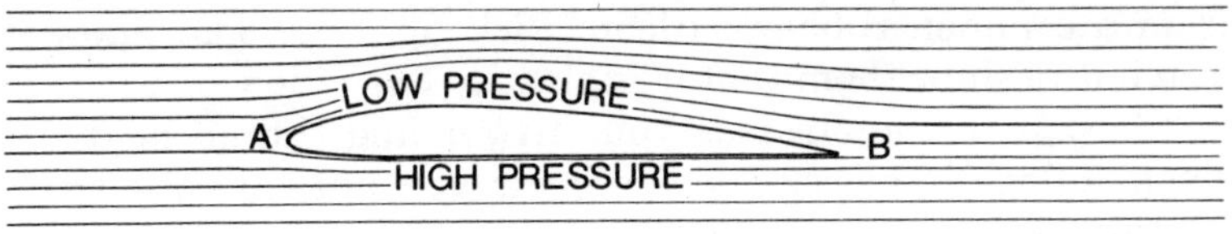

Figure 5-3. Wing moving through air producing lift.

Lift is the force produced by the shape and position of a body moving through a fluid that is directed opposite to the force of gravity (see Figure 5–3). An object moving in a vacuum produces no lift. Lift is generated by an object moving through a fluid if the object's shape causes appropriate reactions as the object moves through a fluid.

ROCKET THRUST

★ Whenever one body exerts a force on another, the second body exerts a force equal in magnitude and opposite in direction on the first body.

What makes a rocket go? It isn't too hard to memorize a set of words from some reference book. Getting an idea of what the words mean isn't too difficult, either. *Understanding* what the words mean or getting the concept isn't always easy. However, it is important that students understand.

What is a rocket? *Webster's Seventh New Collegiate Dictionary* gives the following as one definition for a rocket: "a jet engine that operates on the same principle as the firework rocket, consists essentially of a combustion chamber and an exhaust nozzle, carries either liquid or solid propellants which provide the fuel and oxygen needed for combustion and thus make the engine independent of the oxygen of the air, and is used especially for the propulsion of a missile (as a bomb or shell) or a vehicle (as an airplane)."

This definition is correct, but it doesn't help you understand what a rocket *is*. Perhaps it is best to consider the entire vehicle *and* its engine as the rocket. The distinctive factor about "rockets" is the type of engine that propels them. While there are several types of rocket engines, they all operate on the same principle. A rocket engine moves the vehicle in which it is located by the reaction of the gases that escape from the nozzle of the rocket engine. A rocket engine is a type of reaction engine.

The thrust (push or pressure) developed by a rocket engine is determined by the mass of the gases ejected from the nozzle and the velocity of these gases. The greater the mass of these gases ejected in a given period of time, the greater is the thrust developed. The faster these ejected gases are moving, the greater is the thrust developed.

A good demonstration or hands-on experiment is to make a rocket engine from a common rubber balloon. Pick up a balloon. Show it to the class. Stretch it to show them that it is elastic. Then blow into it to partially inflate it. Show it to the class as your finger and thumb hold the nozzle closed. Then inflate it more. The more air you put into the balloon, the larger it becomes.

Now is a great time for some incidental learning. The next few topics are not essential to understanding why a rocket engine operates, but are relevant. As you hold the partially inflated balloon up for the class to see, ask them to think carefully about the forces holding the air trapped in the balloon. Ask them to ignore the fact that your finger and thumb are holding the nozzle closed. You could eliminate this factor by tying a knot in the nozzle, but this would shortly require untying the knot, a slow and difficult process. Ask the class if they can use reason to determine the two major forces keeping the air trapped inside the balloon.

What are the two major forces acting to keep the air trapped inside the balloon? If you want, ask for volunteers to each supply one of the answers. After all, you have asked the class to try to reach the answers by reasoning. The first answer that will be volunteered probably is "the rubber skin of the balloon." If no one produces this answer, gently rub a finger tip along the surface of the balloon. This produces a loud noise and usually is enough of a clue to prompt the answer. This is a valid answer.

Now ask for the other chief force keeping the air confined in the balloon. Often this produces a long pause, possibly followed by no answers. If you wish, ask the students to take a deep breath and think hard. As they do this, tell them to hold their breath and think hard and that you are certain that the answer is in them. Usually, about the time you start getting a few students with slightly bluish complexions, someone will "explode" with a big sigh and the word "air." That's right, air is all around the balloon. The pressure of the air piled up for hundreds of miles above the balloon produces a significant force on the surface of the balloon.

If you want, you can develop this subject further, either now or later. The actual "weight" of air pushing on each square centimeter of the surface of the balloon can be determined or told. The effect of taking the balloon up onto the roof of the building can be reached by reason (assume that the temperature of the air in the balloon and around the balloon remains the same). The balloon will become slightly larger because of the reduced pressure of the air surrounding it. The effect of taking the balloon down into the basement, assuming the air temperatures remain the same, can also be guessed. The balloon becomes slightly smaller because of the increased air pressure pushing on it.

The effect of adding more air to the balloon can be demonstrated and discussed. The effects of this at the molecular level can also be discussed. One important point that should be made is that the molecules of air in the balloon are bouncing off the inside of the balloon. This produces a pressure

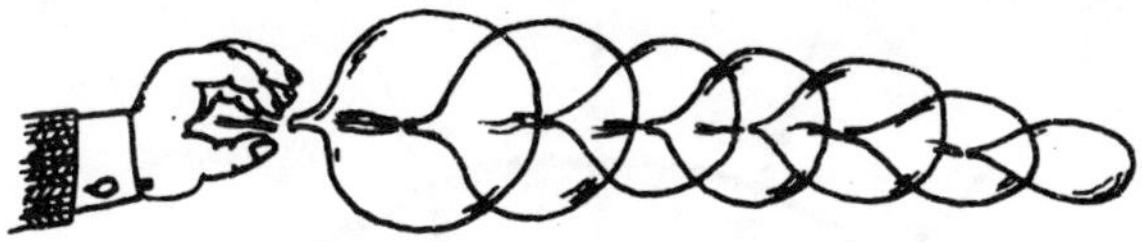

Figure 5-4. Motion of balloon after release.

on the inside of the balloon at every point of its interior. This seemingly small point is important for later discussion.

Now ask the students what they predict will happen, based on their previous experiences, if you release the nozzle of the balloon. Once each one has a mental image of what they expect to happen, release the balloon. Its flight will probably be fairly short and erratic. Retrieve the balloon and again partially inflate it. Apologize for the ineffective guidance system but say that you want to try to achieve a better flight this time. Again blow into the balloon to inflate it even more than you did for the previous flight. Perhaps select a "target" that you wish to reach on the back wall of the room. Release the balloon (see Figure 5–4). Retrieve it and review as you think suitable.

Now tell the class that you want to make sure that everyone understands what has happened. Again inflate the balloon. This time tell the students that you are going to review what happened step by step. As you let a small stream of air out of the balloon, point out that air is rushing out of the nozzle of the balloon. They can't see the air. They can't smell it. They can see that the balloon is getting smaller. Perhaps they can hear a noise that they attribute to the air escaping from the balloon.

Let the stream of air escaping from the nozzle be directed at your hair and allow the students to see it gently blowing your hair around. Or you can let the air stream blow on a few threads or light streamers or paper while the class watches. Point out that the air rushing out of the balloon is displacing the air near the nozzle and any light objects that it strikes.

Review orally and make sure that the students follow your reasoning. Talk to the students about what happened using the method outlined here. The air in the balloon is compressed. The air molecules are tightly packed together. The molecules of air inside the balloon are kept confined by the ballon itself and the air surrounding the balloon. The molecules of air inside the balloon produce a pressure everywhere inside the balloon. As you partially release the nozzle, a small stream of air rushes out of the nozzle. These air molecules push against the molecules of air they strike. So, the air rushing out of the balloon pushes on the air near the nozzle, and this, in effect, pushes the balloon forward (see Figure 5–5). After your group agrees with you this far, tell them that this is *not* correct!

If the molecules of hot gases rushing out of the nozzles of a Saturn V's engines had to have air to push against, how was the Saturn V able to take men to the moon when there is essentially no air for the rocket engine's

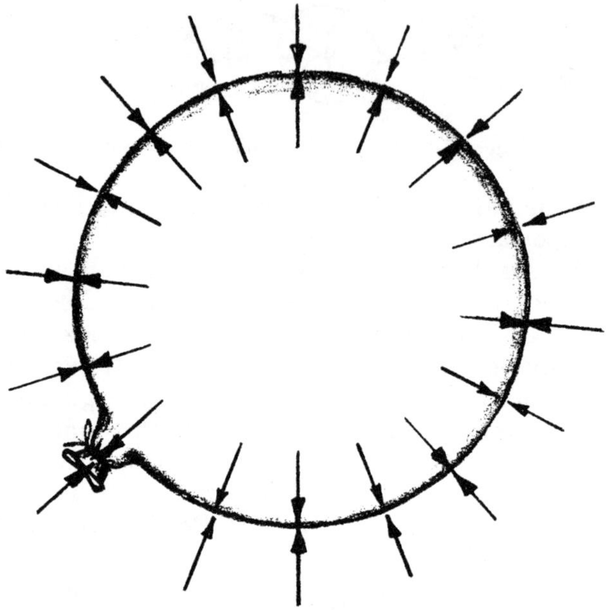

Figure 5-5. Forces upon inflated balloon.

exhaust to push against for most of the trip to the moon? Pause to let this sink in. Discussion at this point can be valuable, but it can also lead to some additional misconceptions and can lead to "defensive thinking" on the part of some students as they attempt to defend their own first opinions.

Again review that there is a force pressing on the inside of the balloon wall at the "top" of the balloon as you hold it horizontally (with the nozzle to one side). Actually point to the locations as you talk about them. Ask if there is a force pushing outward inside the balloon at the "bottom" of the balloon. Of course. Point to a few more locations with the same question. Then point to the very front of the balloon and ask the same question. Then point to the rear of the balloon and repeat the question. Relax the pressure on the nozzle slightly to again let out some of the air in the balloon.

Now ask the students if releasing the nozzle lets part of the air escape from the nozzle. Does this release some of the pressure on this area? Now that this pressure is released by the escaping air, has this removed the pressure against the inside front surface of the balloon, opposite the nozzle? No. The unbalanced force of the air pushing against the front interior surface of the balloon pushes the balloon forward. The air released by the nozzle, just like the hot gases leaving a rocket engine's nozzle, are "garbage" once they leave the nozzle. The unbalanced force on the inside front end of the balloon, or the rocket engine, is what pushes the rocket through space. Perhaps illustrating the balloon on the board with "force arrows" will help the students understand this concept. Any student who has fired a shotgun or a large-caliber rifle will be familiar with the "kick" produced when a bullet is fired.

It is certainly possible to present the concept of the force pushing the balloon or a rocket engine forward in a simpler manner and without so many "incidental" comments. However, one of the authors has used modifications of this approach with success with audiences from middle-school age to adults. Pace your presentation as you watch the reactions of your audience. Keep in mind the main points you wish to present. No matter how well you think you have presented the material, you have failed to educate if the audience has not learned.

One additional quip that has often proved effective is the comment, as you repeatedly blow up a balloon, that you're "a real expert on hot air"!

Another "grabber" to keep the attention of your students (this "lesson" can easily take fifteen minutes, a long time for some students to keep their attention focused) is to slowly keep on blowing up one of the balloons and keep on blowing it up. . . . This is especially effective if you stand very close to one row of students as you do this. Keep talking and keep blowing. Very soon you may notice some cases of "itchy eardrums" or other symptoms of nervousness. Act surprised, and ask what is the matter. When someone expresses concern about the imminent "death" of your balloon, act surprised. Ask why they are concerned. Don't they trust American technology? Do they suspect that each balloon did not receive a good quality check before it was packed? Is it possible that, even with the best of quality checks, which this balloon probably did not receive, that there is a weak spot somewhere on the balloon? Of course there is! Keep on blowing into the balloon. Eventually it will break. Now remark that you have accomplished two objectives: You have invested a dime (or whatever) into the American economy. And you now have the attention of everyone in the class (and perhaps a few rooms down the hall, too)!

A simple review—or even a brief, basic lesson—is to let each student have a small rubber balloon. Let each one blow up the balloon, then let the thrust of the balloon created as the air escapes push the balloon against the student's hand, as shown in Figure 5–6. This is a simple, easy-to-under-

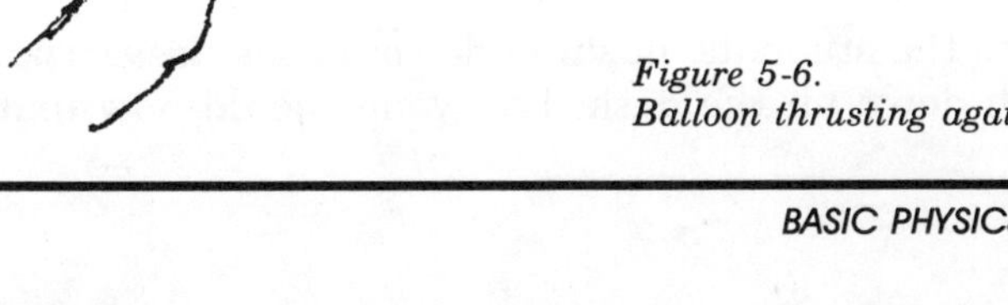

Figure 5-6.
Balloon thrusting against hand.

stand lesson. Each student can perform his or her own experiment. Not only can the student "see" the balloon moving away from the "exhaust" of its nozzle, he can feel the force pushing the balloon against his or her hand.

Now is a good time to review the reason a rocket engine can propel an object through space. You may suggest that a few students still look skeptical about why a rocket engine functions effectively in space. Say that rather than doing another demonstration, you will let the students each perform an individual experiment to better understand the concept of action and reaction. (This experiment may be skipped if you feel that the members of the class already understand the concept of action-reaction.)

Ask the class members to select a small area of the table or desk in front of them. Place an index finger on the spot. Press down gently. Then ask if they are applying a force on the desk. If not, they have not followed directions.

Now ask them to push down harder on the desk. Ask them if they are pushing on the desk. They should realize that they are applying a force on the spot on the desk. Ask them if the desk is pushing back. This may bring about some puzzlement. If you have the time and the inclination, ask for a show of hands of those who believe the desk is pushing back and those who believe that the desk is not pushing back. After the vote, ask them to vote again, and this time look around to see how others are voting. Vote again. You may now ask for individual students to express their opinions. This can be informative, but it can also be very time-consuming and possibly will result in only the more outspoken students expressing their views, well-informed or otherwise. A tactic that has proven productive for one of the authors is to ask the students to have a 30 to 60 second "buzz session" to explain to those near them who voted "wrong" why their reasoning is wrong. Suggest that they convince those who are "wrong" by logic, not by volume.

At the end of the designated time, have the class vote again. Results may not have changed much. As you talk to the class, lean against a wall with one outstretched arm (see Figure 5–7). Ask them to tell you your next two "questions." Tell the class that you are giving them a visual clue. Tell them to look closely at you. Stand so that you are leaning heavily on one hand with the outstretched arm well away from your body and your feet well away from the wall. Stand this way so your center of gravity is not over your feet. The two key questions that should be brought out and answered by the discussion about your visual clue are (1) are you pushing against the wall? and (2) is the wall pushing back against you? The answer to each question is yes. If this were not so, you would fall through or push down the wall ("collapse" your arm so that you "fall" against the wall as the second point is brought out.

Getting back to the students' pushing down on the desk, ask the students to again push down on the desk. Everyone should now understand

Figure 5-7. Teacher leaning stiff-armed against a wall.

that the desk is indeed pushing back against his or her finger. Now ask the students how hard the desk is pushing back against the finger. This question has an answer, although not necessarily a quantitative answer. How hard the desk is pushing back against the finger depends on how hard the finger is pushing on the desk. The desk is pushing back just as hard as the finger is pushing down. Should the students push down harder than the desk is pushing back, they will either push a hole through the desk or break the desk (see Figure 5–8).

Should the desk push back harder than they are pushing down, they have the opportunity to do some first-hand research in magic (see Figure 5–9)!

Figure 5-8 (above). Finger pushing down harder than desk "pushes back."

Figure 5-9 (right). Table "pushing back" harder than the applied force.

The principle of action-reaction can be stated as follows: "Whenever one body exerts a force on another body, the second body exerts a force equal in magnitude and opposite in direction of the first body." This is Newton's Third Law of Motion. A more detailed explanation of this law in terms that your students can understand is given in the article "What Rockets Push Against in Space," the third article in the Estes Industries publication entitled *The Laws of Motion and Model Rocketry.*

THRUST MEASUREMENT

★ The thrust developed by a rocket engine is determined by the mass of gases ejected and the velocity of these gases.

The formula for determining this thrust is:

$F = m\,a$
F = Force
m = Mass of the gases
a = Acceleration of the gases

The thrust developed by a model rocket engine is measured in Newtons. The thrust of each type of model rocket engine is given in graphs published in the catalogs of model rocket engine manufacturers. A time-thrust curve is a way to graphically represent the way a model rocket engine produces thrust. (See Figure 5–10 for the curve for Estes Model rocket engines).

Analysis of the time/thrust curves of model rocket engines reveals several interesting facts. Most model rocket engines have a dual thrust level. All engines have a high thrust level just after ignition. Model rocket engines are made this way to provide a little extra thrust to get them safely started in stable flight. This high initial thrust level is created by the engine having a slight hole in the propellant adjacent to the nozzle. This design allows easier ignition and a high thrust level to quickly accelerate the rocket to a safe flying speed. After this initial high thrust, the thrust of most engines drops to a much lower level, which is usually constant for the duration of the "burn" (the time the engine is burning). This lower, sustained level of thrust is called the sustained thrust level.

A few model rocket engines are designed for only a quick, strong thrust with no sustained burn. These engines produce a "spike" thrust pattern, as compared to the typical engine, which has a spike and a sustained burn pattern. These engines are especially useful for lifting heavy payloads and for other uses that require very rapid acceleration.

Figure 5-10. Comparative time/thrust curves of Estes model rocket engines.

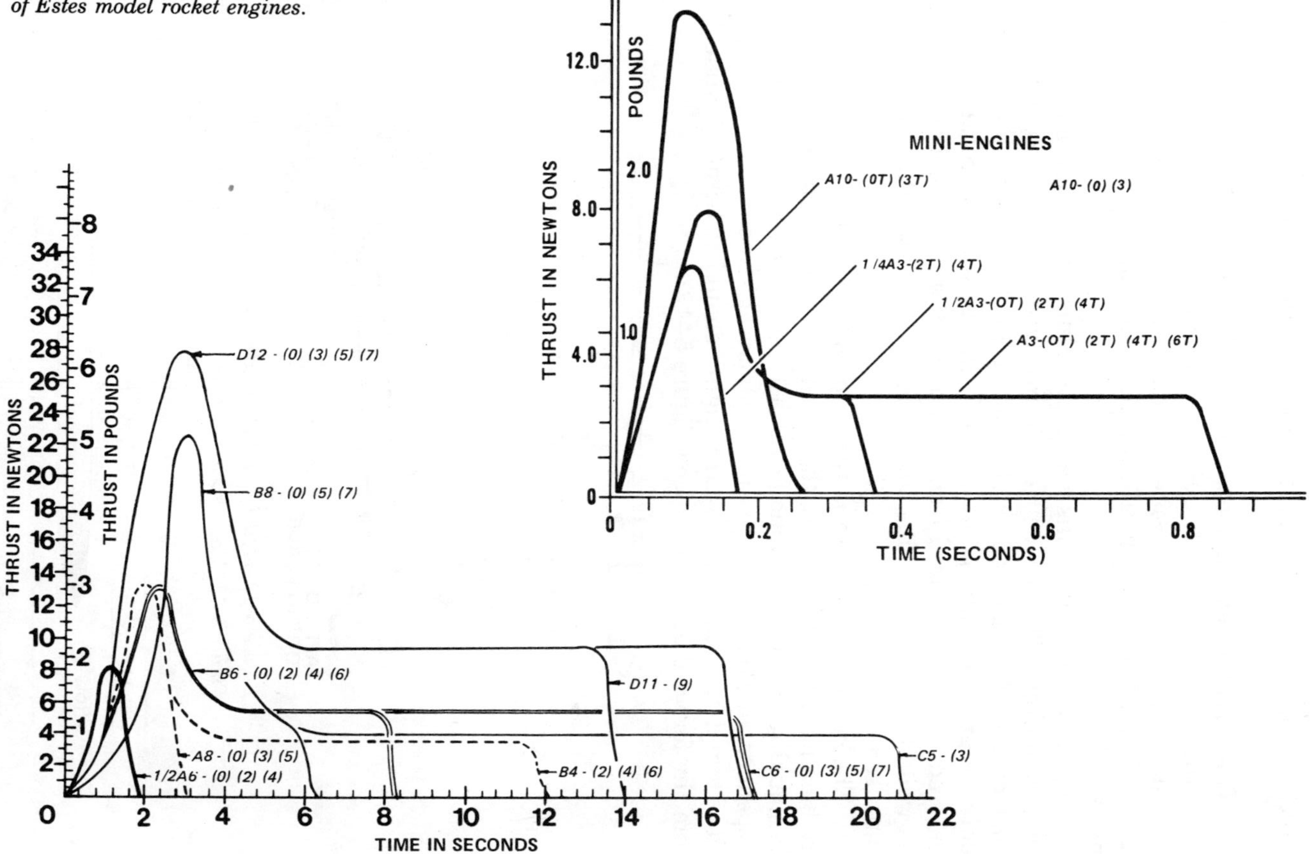

ROCKET GUIDANCE

★ A rocket needs a guidance system to direct its flight.

The early rockets developed by the Chinese relied on a long stick tied to the rocket for their crude guidance (see Figure 5–11).

As the use of rockets spread, modifications to the original design occurred. One of the most significant improvements was the development of spin stabilization.

The development of fins was another step in the development of guidance for rockets. The basic principle involved is that the center of gravity must be ahead of the center of pressure for the rocket to be stable. The center of gravity is the point at which the mass of the rocket seems to be centered. When suspended from this point, the rocket is balanced because the weight forward from this point is equal to the weight to the rear of this point (see Figure 5–12).

The center of pressure is the point where all the aerodynamic forces are balanced. In other words, the aerodynamic forces forward of this point are equal to the aerodynamic forces behind this point. If a model rocket is suspended at its center of pressure in a moving stream of air, the rocket will not attempt to return to "foward flight" should it become "aimed" away from the forward orientation (see Figure 5–13).

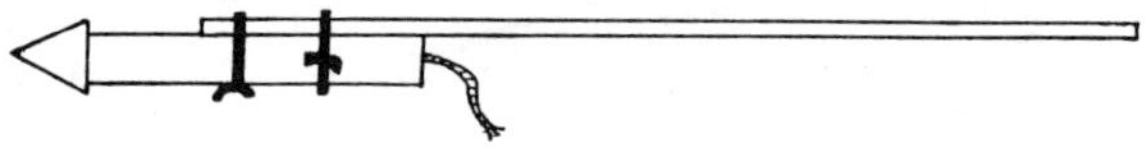

Figure 5-11. Old Chinese rocket attached to stick.

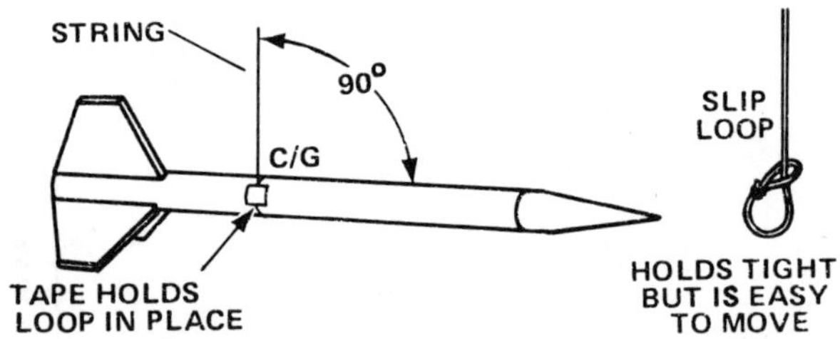

Figure 5-12. Model rocket suspended at its center of gravity.

Figure 5-13. Model rocket held on holder in stream of air.

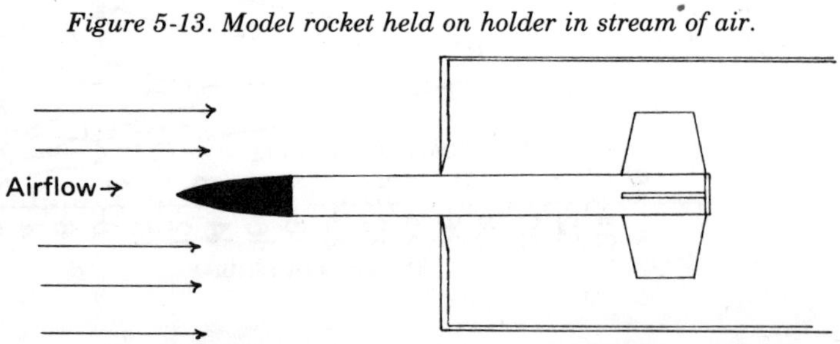

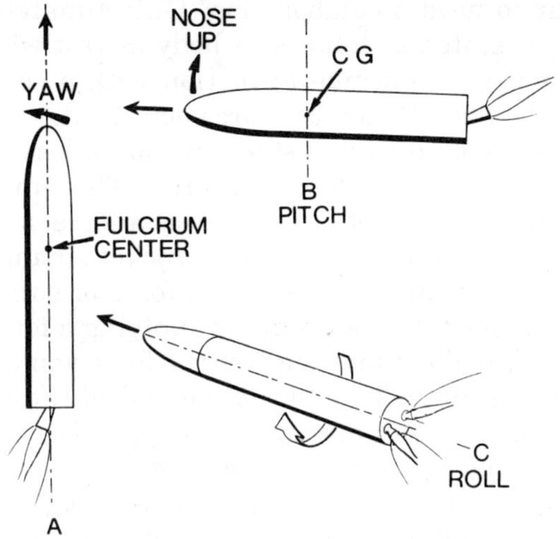

Figure 5-14. Rocket with engine gimballed.

To guide itself in a forward direction while moving through the air, the rocket must have its center of gravity forward of its center of pressure. For model rockets, the distance between the center of gravity and the center of pressure must be at least one-half caliber (half of the diameter of the body tube). Generally, the greater this distance, the more stable the rocket. If the separation of these two points is less than this distance, the rocket may be unstable. Unstable rockets will not return to their original flight path if disturbed from this path while in motion.

If a rocket is to operate in space, it cannot rely on fins for guidance. Fins need air to guide effectively. In space there must be another method for guidance of the rocket. One of the most common methods of directing the rocket while in motion in space is the use of gimballed engines (see Figure 5–14). These permit the direction in which the engine is thrusting to be changed, thus changing the direction of motion of the rocket.

Another method for guiding a rocket without the use of fins is the use of deflector vanes in the rocket's exhaust. These vanes cause the direction of the rocket's exhaust to be directed away from the rocket's main axis.

LAWS OF MOTION AND MODEL ROCKETRY

★ Model rockets obey the same principles of physics as their full-sized counterparts.

The Third Law of Motion, often referred to as the law of action and reaction, has been presented earlier in this chapter.

The First Law of Motion, often referred to as the law of inertia, is so basic it may appear to need no elaboration. Unfortunately, this is not the case. This law can be stated as follows: A body at rest will remain at rest, and a body in motion will continue in motion with a constant speed in a straight line so long as no unbalanced force acts on it.

This may be considered in two steps. A motionless body will remain motionless unless an unbalanced force acts on it. This simply means that a stationary object will remain motionless unless a force of sufficient magnitude acts on it. Take the example of a model rocket sitting motionless on a table. The rocket will remain there unless a force of sufficient magnitude acts on it. If you try to pick the rocket up by applying a force vertically with your hands, the rocket will not move unless the force you apply is enough to overcome the force of gravity. The "static inertia" of the rocket keeps it in place unless a force of sufficient magnitude is applied to cause the model to move. Most students find this concept easy to accept.

The inertia possessed by a moving body is sometimes referred to as "kinetic inertia." The body will continue to move in the same direction at the same speed or velocity unless acted upon by some unbalanced force. In space, this means that a moving object tends to keep on moving in the same direction at the same speed unless an unbalanced force acts on it (see Figure 5–15).

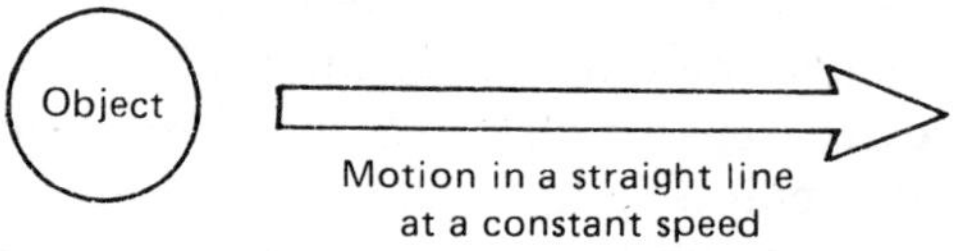

Figure 5-15. Object in motion in space.

Gravity is a force that acts upon objects near each other in space. The larger the objects, the greater the force with which they are attracted toward each other. The closer together the objects are to each other, the greater the force with which they are attracted toward each other. A space ship moving through space will tend to keep on moving in the same direction at the same speed unless acted upon by an unbalanced force. Such forces can include the gravitational attraction between the space ship and nearby planets or stars and the drag produced by the space ship as it moves through the gases and dust particles that may occur in space.

One relatively simple way to quantitatively look at the energy possessed by a moving object is to determine the *momentum* of the object. Momentum is determined by multiplying mass times velocity. A stationary object possesses zero momentum (no velocity). The momentum possessed by a moving object varies as the velocity of that object changes. For example, the momentum of a model rocket increases as the velocity of the model rocket increases. However, the mass of a model rocket is reduced slightly as the propellant (fuel and oxidizer) of the model rocket engine is reduced as this mass in the engine is converted to hot gases and ejected from the engine's nozzle.

For a more detailed look at the First Law of Motion in terms that are understandable to junior high-school students and includes a demonstration, please read "Why Satellites Stay in Orbit," the first article in *The Laws of Motion and Model Rocketry,* published by Estes Industries.

The Second Law of Motion is often harder for youngsters to understand than the other two of Newton's Three Laws of Motion. The wording is longer, and the concepts are more complex. Students can, however, grasp the concepts and use them. The Second Law of Motion may be stated as follows: If an unbalanced force acts on a body, the body will be accelerated; the magnitude of the acceleration is proportional to the magnitude of the unbalanced force, and the direction of the acceleration is in the direction of the unbalanced force.

Perhaps the best way to understand these concepts is by breaking down the law into "bite-size pieces." The terms used must be clearly understood if the concepts are to be understood. An "unbalanced force" is one that is not matched or balanced by an opposing force. An acceleration is a change in speed. Magnitude refers to size or quantity.

The Second Law of Motion is sometimes referred to as the law of momentum change. Remember that momentum usually refers to the product of mass times velocity. One simple demonstration of a change in momentum is to simply drop a small ball. The ball rapidly gains speed as it starts to fall from your hand. The ball falls because of the unbalanced force of gravity acting on it. The ball is actually accelerating as it falls. This means that the ball gains momentum as it falls. The mass stays the same, but the velocity of the ball changes. However, the ball will not accelerate indefinitely, even if released from an airplane high in the sky so that it can fall for a long time before striking the surface of the ground.

The ball is passing through air. This air resists the movement of the ball through it. This resistance is called drag. Thus, the ball is subject to acceleration toward the ground because of gravity, and the ball is prevented from accelerating indefinitely, even before it strikes the ground, because of the drag of the air. The ball will continue accelerating as it falls until it reaches a speed at which the drag of the air balances the acceleration due to gravity. This final, maximum speed is called the terminal velocity of the ball.

Further study of the forces acting upon a ball can be informative. If the ball is tossed vertically into the air, does it continue going up indefinitely? No. As the ball leaves your hand, it achieves a certain velocity the instant it leaves your hand. This is the maximum upward velocity of the ball. As the ball rises, its motion is resisted by the air producing drag, which slows the upward motion of the ball. You can call this negative acceleration, because the ball is slowing down. The ball is also being attracted toward the center of the earth by gravity, so the force of gravity is acting on the ball to slow it down. This is another force producing negative acceleration. As a result of these two forces, the ball soon slows down and stops. At this instant the ball has zero momentum because it has zero velocity. The ball is subject to gravity, however, so it starts to accelerate downward. This positive acceleration (increase in speed) is resisted by the drag that the ball encounters as it moves through the air.

One of the authors has found the previous series of demonstrations with a ball to be quite effective in capturing and retaining the interest of students and in helping them understand the effects of the forces of gravity and drag. In fact, the study of these simple demonstrations can be effectively extended. For an older class, toss up the ball vertically and catch it again, with the height from which you release the ball as close as possible to the height at which you catch the ball. Ask the students to observe carefully and then ask them to close their eyes and recreate the series of events as an "instant replay."

Now ask them to tell you the exact point at which the ball had maximum velocity. Let several students answer. Handle the answers appropriately. The point you are trying to make is that the ball actually has maximum velocity at *two* points, the point at which you released it as you tossed it up and the point at which you caught it.

Depending upon the ages of your students, the depth of your prior coverage, and your own goals, this activity can be carried further. For older students, ask them to describe the motion of the ball through its flight or trajectory. They should realize that the ball is slowing down (negative acceleration) on the way up and gaining speed (positive acceleration) on the way down. Ask them if the ball actually stops. This question is sometimes a good one to introduce before you have finished examining the motion of the ball in its flight. You can give it as a homework assignment. Let students work in pairs, with one tossing up the ball and catching it while the other observes, then reversing roles. You may have to state in the discussion the following day that you are ignoring the spin of the ball and that the ball probably did not go up or fall exactly vertically.

Did the ball actually stop? The answer can be either yes or no, depending upon the viewpoint. This is not a contradiction, but rather a way of looking at things clearly. As far as you are concerned, the ball does stop at the top of its flight. In actuality, the ball is moving with the earth as it rotates, moving with the earth as it moves in its orbit around the sun, and so on. In one "frame of reference," the ball stopped. In another, it did not stop.

Another good "thought" question to ask the students, possibly on a different day, is based on your selecting a suitable "target," or making one, at the back of the room, preferably at about your chest height when standing. Stand at a convenient location near the front of the room and ask the students "How many tries will it take for me to throw the ball straight to that target at the back of the room?" Make certain that they see the target. It adds a certain tension if the target is rather close to several students. Don't worry! You are not actually going to throw the ball at the target, but the students do not know this. Let students volunteer guesses. Answers may range from one on up. Repeat the question several times exactly as stated as you ask different individuals. Keep asking until someone answers that you will never do it. Ask this alert person to explain his or her lack of faith in your pitching prowess. The point you are trying to get across is that you cannot throw the ball *straight* to the target. You may hit the target on the first try, but you will do it by throwing a curve. The ball must be thrown slightly up to arc down to the target because the ball will be pulled downward by the force of gravity the entire time it is in the air.

Perhaps these digressions are not essential to get the concepts across that you are trying to emphasize, but they are helpful in getting students to reason. Quite often we need to ask the right question or listen carefully to the facts to gain a better understanding. The old expression "It isn't what we don't know that gets us into trouble. It's what we know that isn't so!" is still quite true.

The use of staged rockets provides a good example of changing the speed of a body by applying repeated unbalanced forces to it. Each added application of thrust accelerates the body in the direction of the thrust. A simple example that is very useful with younger students is to line the students up along a hall or in a gym. Gently roll a large ball, such as a basketball, along the line of students. If each student hits the ball in the direction in which it is rolling, each provides positive acceleration to the ball. If a student hits the ball harder, the ball moves faster. If a student hits the ball in the direction opposite its motion, however, the ball is slowed down. If the momentum of the ball is greater than the momentum of the blow striking it, the ball slows down but does not stop. If the force of the blow striking the ball is the same as the momentum of the ball, the ball will stop, except for rebounding because of its elasticity. If the force of the blow is greater than the momentum of the ball, the ball will stop and be forced backward.

Staged rockets use a series of unbalanced forces acting upon the rocket. See Figure 5–16. Each unbalanced force increases the velocity of the rocket. As each "burn" takes place, the rocket is accelerated so that its velocity increases. However, as each "burn" occurs, the mass of fuel and oxidizer within the rocket is reduced. The mass of the expended stage (engine casing, fins, and so on) is also removed at staging. This reduces the mass of the rocket. This means that the mass of the rocket is less for each subsequent "burn." The less the mass that is moved, the greater the velocity that a

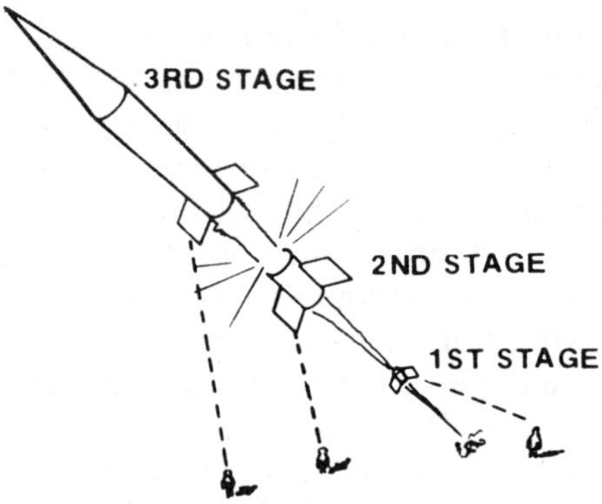

Figure 5-16. Staged rocket.

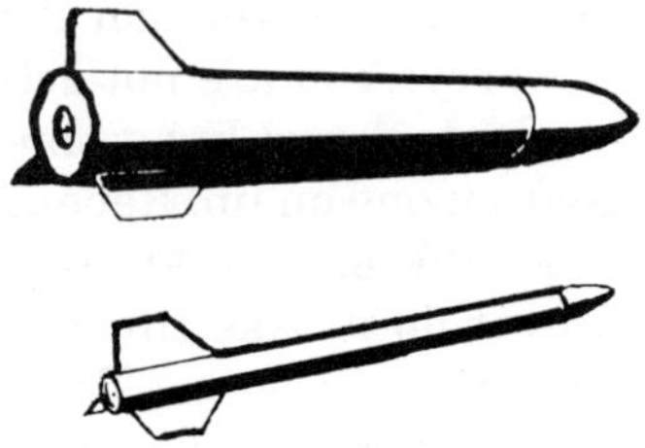

Figure 5-17. Effects of a specific thrust on two rockets of different masses.

given force will produce. In other words, a certain amount of thrust will produce a greater acceleration in a rocket of small mass than in a rocket of great mass, as shown in Figure 5–17.

The Second Law of Motion is covered in a slightly different form in "Why Rockets Are Staged," from *The Laws of Motion and Model Rocketry.*

ORBITAL MOTION

★ Your model rocket won't make it to orbit.

Let's spend a little time studying orbital forces, however, because rocketry is closely related to satellites. The only way yet discovered to put an artificial satellite into orbit is with rocket power.

To achieve a better understanding of relative speeds and the speed of a satellite in orbit, let's consider speeds of several common objects.

Speeds are usually expressed as average speeds. To determine average speed, add the initial speed of an object and its final speed, then divide the total by two. If a car starts from a stationary start and slowly and steadily accelerates to reach a speed of 50 miles per hour after it has been in motion for one hour, what is its average speed? The original speed of 0 miles per hour plus its final speed of 50 miles per hour divided by two yields an average speed of 25 miles per hour. This car will travel 25 miles in its first hour of travel.

A man walking at a steady pace averages about 3 miles per hour. This is about 4.4 feet per second. A fast racer can run about 100 yards in 10 seconds. This is about 30 feet per second or 20 miles per hour.

To compare relative speeds, it takes about

- 68 seconds to walk 300 feet (length of average city block)

- 10 seconds to run 300 feet
- 0.45 seconds for a fast model rocket to travel 300 feet
- 0.27 seconds for a jet plan flying at the speed of sound to travel 300 feet
- 0.01 seconds for a satellite in a 100-mile-high orbit to go 300 feet.

FALLING OBJECTS

★ A falling object accelerates (changes speed) at the rate of 32 feet per second each second it is falling when it is near the surface of Earth and not influenced by other forces. This means that a ball dropped will travel 16 feet in its first second of fall. The initial speed of 0 feet per second plus its speed at the end of one second of 32 feet per second divided by two yields 16 feet for the distance traveled.

Using this same method, how far will a ball fall in its first two seconds of fall?

$$\text{average speed} = \frac{\text{original speed} + \text{final speed}}{2}$$

$$= \frac{0 \text{ feet per second} + 64 \text{ feet per second}}{2}$$

$$= 32 \text{ feet per second}$$

$$\text{distance traveled} = \text{average speed} \times \text{time in motion}$$

$$= 32 \text{ feet per second} \times 2 \text{ seconds}$$

$$= 64 \text{ feet}$$

How far does an object fall in its third second of fall?

To make solving such problems easier, let's take a few minutes to restate what we have determined in general equations that can be used for most similar problems. The formula for average velocity can be stated:

$$v = \frac{v_1 + v_2}{2}$$

v = average velocity
v_1 = original velocity
v_2 = final velocity

The formula for determining the distance an object falls during a given time is:

$$s = \tfrac{1}{2} g t^2$$

s distance
g acceleration due to gravity
t time

To solve the problem, calculate the distance the object falls during three seconds, then subtract the distance the object falls in two seconds. While there may be other ways to determine this distance, this method works.

$$
\begin{aligned}
s &= \tfrac{1}{2}\, g\, t^2 \\
&= \tfrac{1}{2} \times 32 \text{ feet per second}^2 \times (3 \text{ seconds})^2 \\
&= \tfrac{1}{2} \times 32 \text{ feet per second}^2 \times 9 \text{ seconds}^2 \\
&= \tfrac{1}{2} \times 32 \text{ feet} \times 9 \\
&= 16 \text{ feet} \times 9 \\
&= 144 \text{ feet}
\end{aligned}
$$

Since the object fell 64 feet in two seconds, 144 feet minus 64 feet is 80 feet. The object fell 80 feet during the third second of fall.

ORBITS

The farther an object is from the surface of Earth, the slower it falls. An object near the surface of Earth falls about 16 feet in one second. Earth's surface "curves down" 16 feet in about 5 miles. An object traveling horizontally at about 5 miles per second will fall at a rate that keeps it the same distance above Earth's surface. Since Earth's atmosphere is very dense at this level, drag will be very great, so it is not practical (for this and other reasons) to place an object in orbit near the surface of Earth. (See Figure 5–18). This speed of 5 miles per second is about 18,000 miles per hour (5 miles/second × 60 seconds/minute × 60 minutes/hour).

Figure 5-18. Moon's position at a given moment.

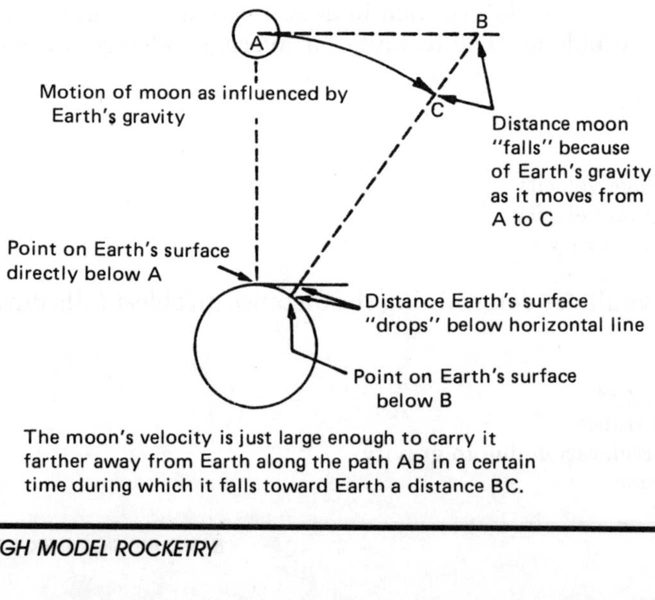

The moon's velocity is just large enough to carry it farther away from Earth along the path AB in a certain time during which it falls toward Earth a distance BC.

Figure 5-19. Velocities and periods of satellites in orbit.

Altitude Miles	*Velocity Miles per sec.*	*Period*
0	4.92	1 hr. 24 min.
100	4.85	1 hr. 28 min.
400	4.68	1 hr. 38 min.
5,000	3.27	4 hr. 47 min.
22,300	1.91	24 hr. ———

Figure 5-20. Comparative orbits and velocities.

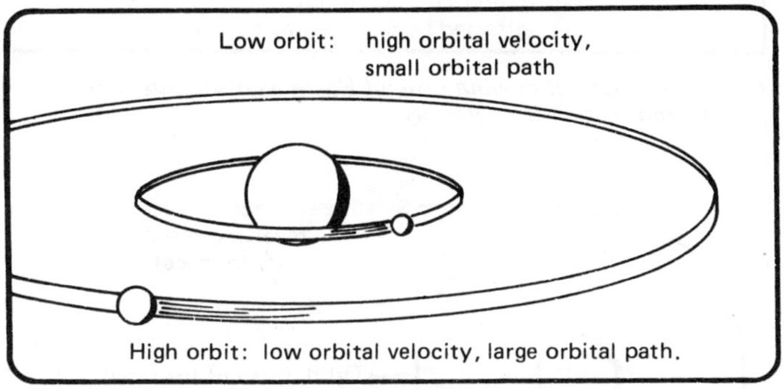

Figure 5-21. Object in motion at constant speed in straight line.

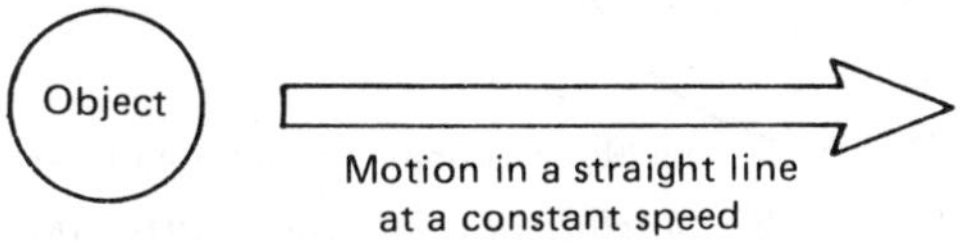

Since the farther an object is from Earth's surface the slower it falls, the slower it has to move to stay in a circular orbit. Satellites very far from the surface of Earth move fairly slowly compared to those in lower orbits. The larger the orbital path and the slower the motion, the longer the period (time required for one complete orbit) of the satellite. (see Figures 5–19 and 5–20.)

The First Law of Motion states "a body in motion will continue in motion with a constant speed in a straight line unless an unbalanced force acts on it." This isn't all that the First Law of Motion states, but it's the part we are concerned with. (See Figure 5–21.)

Objects in space are attracted to other objects in space near them. The strength of this attraction is proportional to the masses of the bodies. The strength of this attraction is inversely proportional to the distance between the bodies. See Figure 5–22.

The object tends to go in a straight line at a uniform speed if not acted

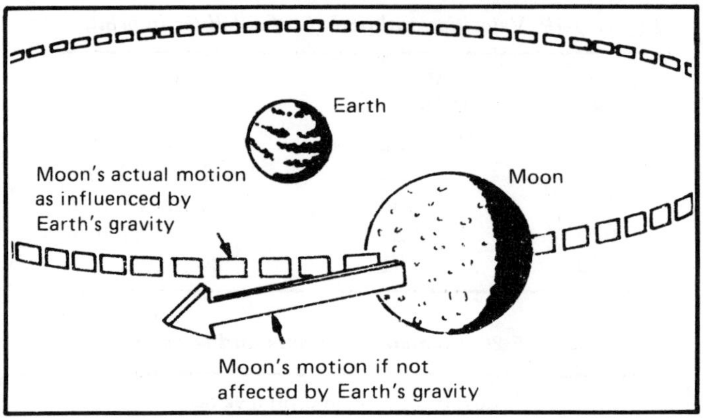

Figure 5-22. Object moving around Earth as a satellite with gravity and motion if no gravity.

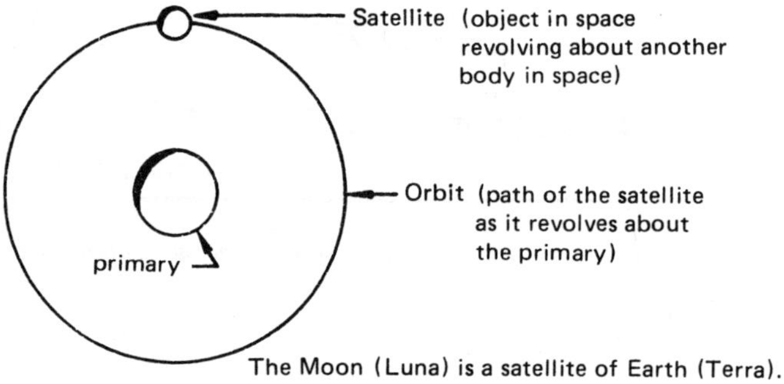

Figure 5-23. Satellite in orbit around primary.

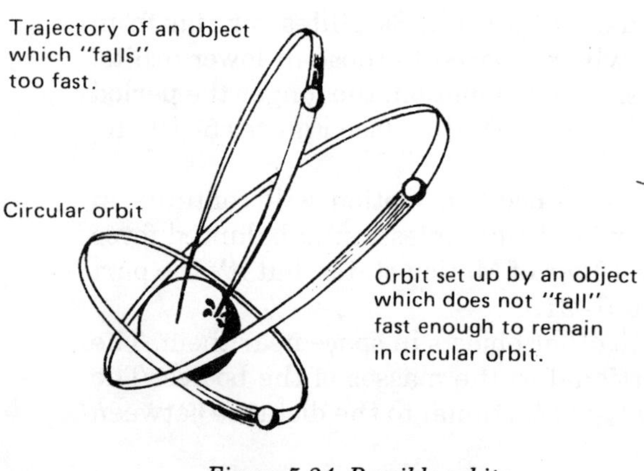

Figure 5-24. Possible orbits.

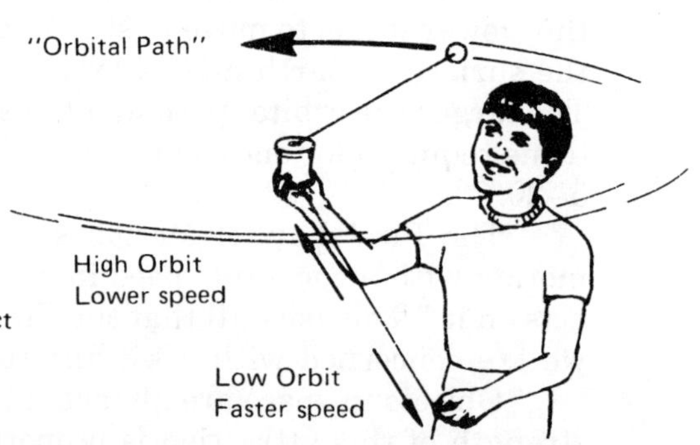

Figure 5-25. Whirling object over head on string.

upon by a force. Since there is a force, gravity, acting on the object, its motion is not in a straight line. This force causes the object to fall toward Earth. Since the object has a lateral velocity, the forces may interact to cause the object to go into orbit (see Figure 5–23).

If the forces acting on the object moving about the primary are not properly "balanced," the object does not go into circular orbit (see Figure 5–24).

The farther an object is from the primary, the weaker the force is with which gravity pulls on the object. This means that the higher an object is above Earth, the slower it falls. This also means that the higher an object is above Earth, the slower its movement is in its orbit. If the object is caused to speed up, its orbit will change. Also, if the object is slowed, its orbit will change.

The higher the orbit is, the slower the satellite's motion in orbit is and the longer its period (time required to complete one orbit) is. This means that a satellite in a low circular orbit has a shorter period as well as a shorter orbital path and a faster speed.

A graphic simulation of orbital motion is easy to set up as a demonstration. Its real value occurs, however, if each student is permitted to try it for himself. Seeing it done is effective, but when the student can actually *feel* the forces involved, the demonstration is much better understood and retained longer. The apparatus needed is simple and easy to secure.

A piece of strong string (about 5 feet long), a fairly lightweight object (as a large nose cone or a rubber ball of about 3 inches in diameter), and a sewing thread spool (may be empty or not) are needed.

Tie the object to one end of the string. If you use a rubber ball, one end of the string may be inserted *carefully* through the ball (as with a small screwdriver) and the end of the string tied securely to the string entering the ball.

Put the string through the spool with about 1 foot of the string on the side of the spool with the object. Make sure that you are in a clear, open area. Holding the spool above your head, whirl the object around and around. Feel the force on the end of the string. Whirl the object faster. Feel the change in the force.

Now let about another foot of the string go through the spool and continue whirling the object (see Figure 5–25). Feel the change in the force. Also notice the change in the speed of the object. Now whirl the object faster and feel the change in the force. Allow the object to slow down and "droop" in its path. Again feel the change in the force.

The object represents a satellite in circular orbit, and the movement of the object represents the satellite as it moves in its orbit. The spool represents Earth. The force on the string represents the gravitational attraction between Earth and the satellite.

While this simulation is not a true replication of the motion of a satellite in orbit, it permits each student who performs it to gain a much better insight into orbital motion and the forces on objects in circular orbits than is usually possible through just reading about the forces. This simulation of "high orbits" and "low orbits" permits each student to better understand the effects of changes in orbital size, speed, and orbital period.

Ask your students what will happen if the string breaks as the object "in orbit" circles about the spool. Of course, the object will tend to fly off into space in a straight line and at the same speed that it was moving when the string broke. The forces of gravity and air drag will soon cause the object to fall to the ground.

AERODYNAMICS

The aerodynamics of flight are very important when considering the forces operating on a model rocket since the entire flight of a model rocket occurs within the atmosphere. These forces will be treated in detail in Chapter 6.

6
AERODYNAMICS AND MODEL ROCKETRY

FORCES ON FLYING OBJECTS

★ The four basic forces on flying objects are gravity, lift, thrust, and drag.

GRAVITY

Gravity pulls on all objects near Earth. The force varies inversely with the square of the distance between the center of gravity of the object and the center of Earth. In other words, object B, which is twice as far from the center of Earth as object A, will experience one-fourth the gravitational attraction experienced by object A.

LIFT

★ Lift is produced if the air flows over the surfaces of the object so that uneven pressures are produced on different sides of the object (see Figure 6–1).

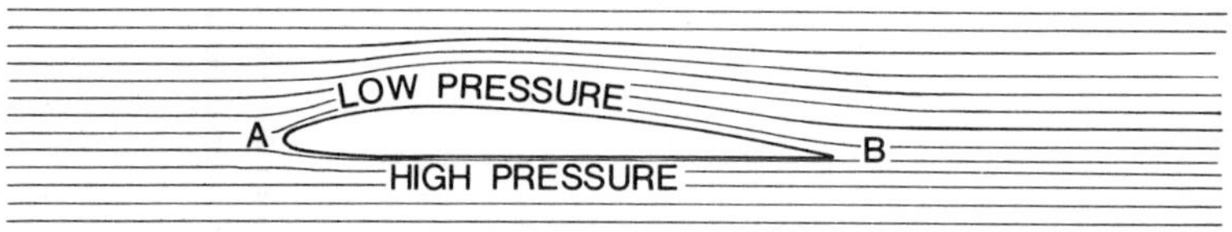

Figure 6-1. Air moving past airplane wing producing lift.

The flight characteristics of several common objects will be examined to make some of the concepts more easily understood.

On a well-designed wing, the air flowing above the wing travels faster than the air flowing below the wing. This increase in speed of the air above the wing produces a decrease in air pressure. The uneven pressures produce a net force, called lift, toward the upper surface of the wing. The greater this difference in pressure between the top and the bottom of the wing, the greater the lift produced. Also, the greater the surface area producing a lift force, the greater the total lift produced.

Lift is very important in the performance of boost gliders. However, lift can be very undesirable for a model rocket during the thrust phase of its flight. Lift can tend to make the rocket fly to one side instead of vertically. If you launch in windy weather, you must allow for the effect of weathercocking to recover the rocket within the intended recovery area.

Kites are simple flying objects. Study of the aerodynamic forces acting upon kites (see Figure 6–2) will help you absorb concepts useful in the analysis of aerodynamic forces acting on gliding bodies, such as a boost glider during the glide phase of its flight.

The angle of attack is the angle between the chord line (the line connecting the center of the leading edge and the center of the trailing edge) of

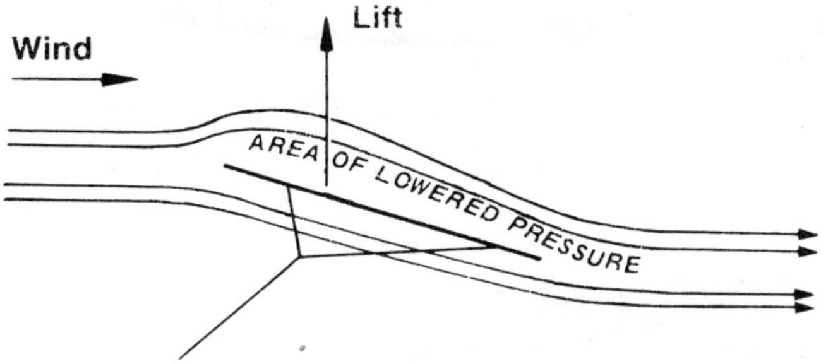

Figure 6-2. Kite flying at angle of attack in relation to wind with differential pressures and lift force shown.

Figure 6-3. Stall.

the wing and the "relative wind." The greater the angle of attack of a wing, the further and faster the air must flow over the wing and the greater the lift force produced. However, if the angle of attack becomes too great, the air flow becomes violently turbulent, and a stall results (see Figure 6–3). This usually results in the loss of the airplane, unless the pilot can cause the plane to recover from the stall. Since your model rocket has no pilot, a stall usually results in a crash.

If the flying object flies at too great an angle of attack, the object will tend to stall. This tendency can be corrected by reducing the angle of attack. This reduction is usually produced by adding weight to the nose of the flying object. The reduction in angle of attack can also be produced by decreasing the weight of the tail of the flying object.

While it may appear that a gliding object is very efficient if it climbs as it glides, it isn't. The object is trading forward speed for height. As it slows, the lift it generates decreases, and it stalls. It may recover from the stall to stall again a few seconds later as it tries to rise. Adding a small amount of weight to the nose or altering the aerodynamic properties of the gliding object can eliminate these stalls. Also, a craft that glides well at low speeds may stall when it attempts to glide at higher speeds.

When the flying object descends too rapidly, the condition is called a dive (see Figure 6-4). If the object cannot recover from the dive, it crashes. Dives may be caused by too much weight toward the nose, too small or inefficient lifting surfaces, too much total weight, or flying speeds that are too low. The angle of attack is too low for successful flight. Adding a small amount of weight to the tail, reducing the total weight, increasing the lifting surface area, improving the lifting surface design, or increasing the flying speed may help reduce diving.

If an object retains a good glide ratio (length traveled horizontally to vertical drop), it will glide well (see Figure 6–5). Stalls and dives don't occur. The glide ratio can be improved by reducing the weight of the object, increasing the lifting surface area, improving the lifting surface design,

Figure 6-4. Dive.

Figure 6-5. Glide.

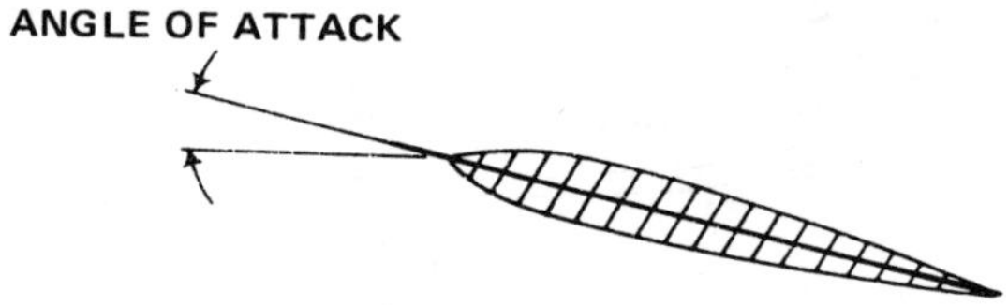

Figure 6-6. Wing chord and angle of attack.

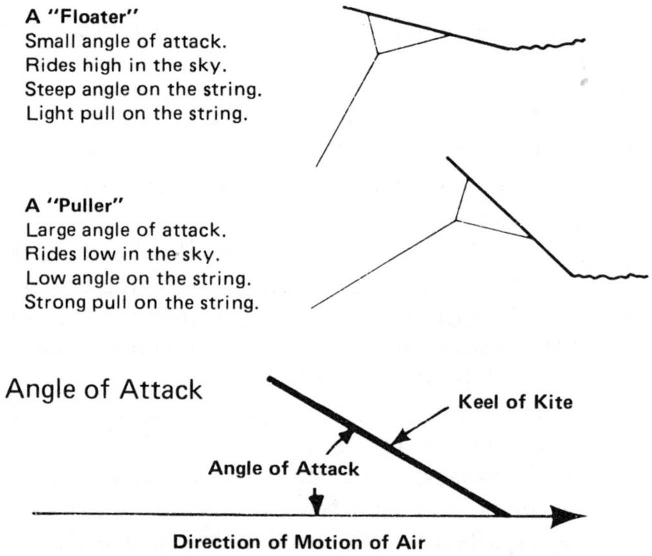

Figure 6-7. A floater kite and a puller kite.

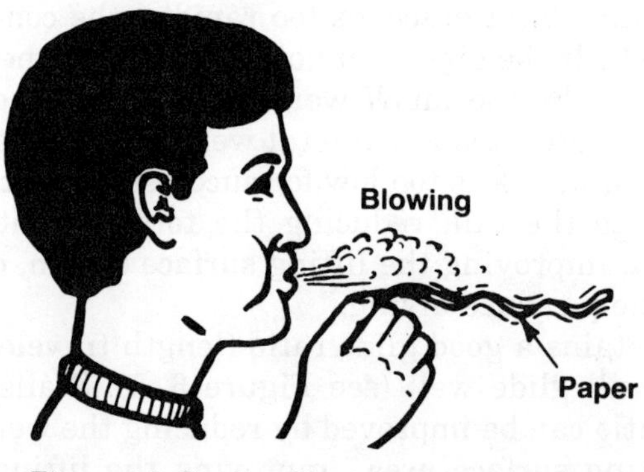

Figure 6-8. Student blowing over a paper strip.

improving the weight distribution (center of gravity to center of pressure relationship), or changing the relative air speed (by increasing the speed of the object, increasing the speed of the wind through which it is moving, or, occasionally, slowing the object's speed) (see Figure 6–6).

Kites with a small angle of attack produce minimal pull on the flying line. These kites are referred to as "floaters."

Kites with a large angle of attack pull hard on the flying line because they have high drag. These kites are referred to as "pullers." See Figure 6–7.

One of the simplest demonstrations of lift is one that your students can perform for themselves. This works best if your students do it as an experiment with no demonstration and very little instruction from you. Provide each of your students with a strip of "limp" paper, such as newspaper, about an inch wide by 6 to 12 inches long. Ask them to hold the paper with their thumbs, with the paper just draping over their index fingers with the long part away from them. Tell them to hold their index fingers near their mouths and blow gently over their index fingers (see Figure 6–8). Watch the paper.

Many students do not expect the paper to rise. This rising of the paper provides a quick and direct demonstration of lift.

Additional information on aerodynamics as they apply to kites and kite flying is available in *Let's Go Fly a Kite!*, third edition, and *Kite Flying Teacher's Guide,* both from Estes Industries.

THRUST

★ Thrust is the propulsive force that moves an object.

Thrust as it applies to model rockets is covered in Chapter 9.

DRAG

★ Drag is the resistance caused by the motion of an object through a fluid.

Drag results from the action of the object moving through the air. The rougher the surface of the object, the greater the friction produced is and thus the greater the drag. The faster the object moves through the air, the greater the drag is. The greater the drag on an object, the more energy the moving object loses to friction rather than producing motion.

Drag may be a difficult concept for some students. Basically, drag is friction.

Any object moving through a fluid interacts with the fluid at all points of contact. The areas of contact are often referred to as the "wetted surface," whether the object is moving through a liquid or a gas. As the object moves through the fluid, the molecules of the fluid must be moved aside. The various forces involved in this movement constitute drag. The energy used in this movement is derived from the energy of the moving body. Much of this energy is converted into heat by the friction of the motion.

One easy way to let your students experience this is to have them rub an index finger gently back and forth across the surface of the desks or tables at which they are seated. At first, nothing is felt by most students except the awareness of the contact between their fingers and the surfaces. Even continuing this motion produces little change in the sensation. Now let the students rub their fingers faster and faster. Soon the students will notice a sensation of heat. Further fast motion intensifies this feeling. Do not let your students continue this motion too long, or a blister may result.

The rougher the surface of the moving object, the greater the drag produced.

Another activity that can be helpful in understanding drag involves your students again rubbing index fingers against a surface. Instead of rubbing faster, have your students rub harder. At first, continuing this motion seems to do nothing special, but the students will soon notice the extra effort involved in moving the finger. While the heat buildup may not happen as fast, the students will notice that the rubbing motion requires more effort and is more tiring, and will also produce heat buildup. It also wears away the surface of the finger and, to a lesser degree, the surface of the desk.

The faster rubbing of the finger against the surface in the first activity can be compared to the object moving faster through the fluid. The faster the motion, the greater the drag produced (see Figure 6–9).

The second activity can be related to an object requiring more force to move through a fluid. Another way to help the students understand this

Figure 6-9. Low drag model rocket.

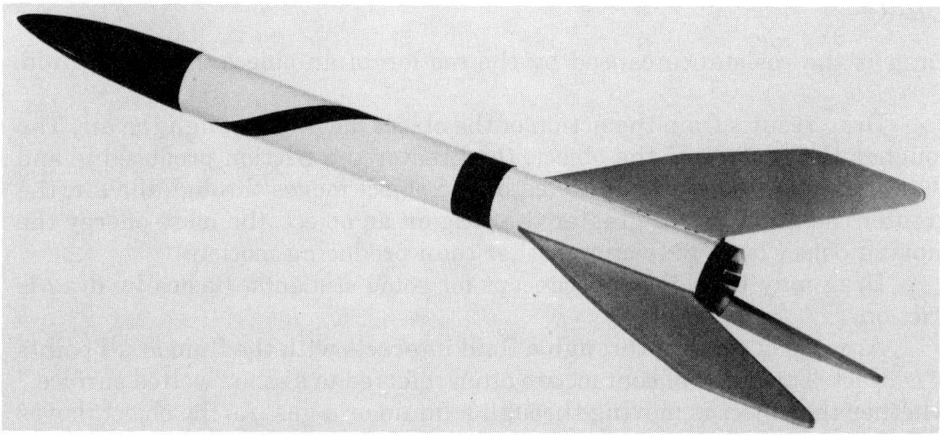

Figure 6-10.
High drag model rocket.

concept is to ask them to think of running fast. They could feel the air rushing past them and the air gently pushing against them, slowing them down. Then ask them to recall trying to run through water that comes up to about their waists. Even at fairly slow speeds, the great drag experienced as they pushed their way through the water was obvious. Water is a much denser fluid than air (see Figure 6–10).

Aerodynamic drag is the resistance produced by the motion of an object through air (from aero, meaning air, and dynamic, meaning forces).

COEFFICIENT OF DRAG

★ The aerodynamic coefficient of drag (C_D) is a measure of how easily a given shape moves through the air. The smaller the C_D number that a shape has, the more efficient the shape is (because less drag is generated in flight) (see Figure 6–11).

LAMINAR AND TURBULENT FLOWS

★ Laminar flow exists when the boundary layer of fluid (air) next to the surface is smooth and "attached" to the surface (see Figure 6–12). Turbulent flow exists when the boundary layer of fluid (air) next to the surface is not smooth and "attached."

In laminar flow, the molecules of air flow smoothly over the surface.

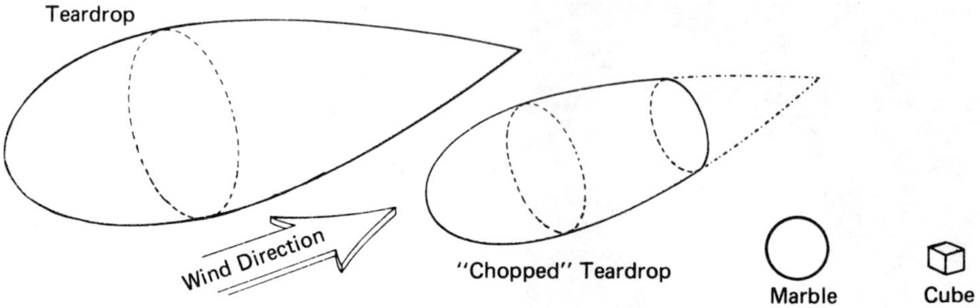

Figure 6-11. Relative C_D of various shapes.

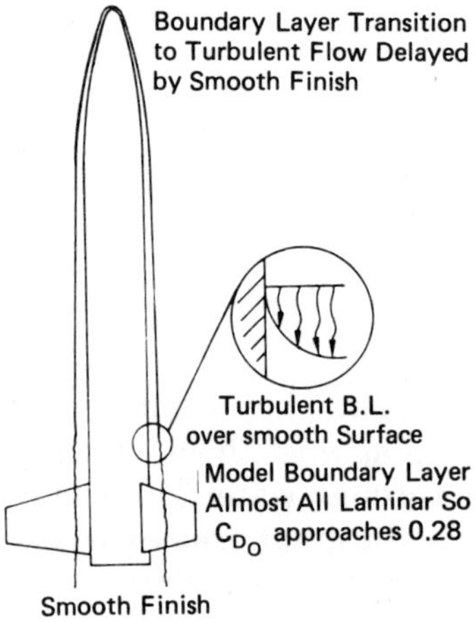

Figure 6-12. Laminar flow over front section of model rocket.

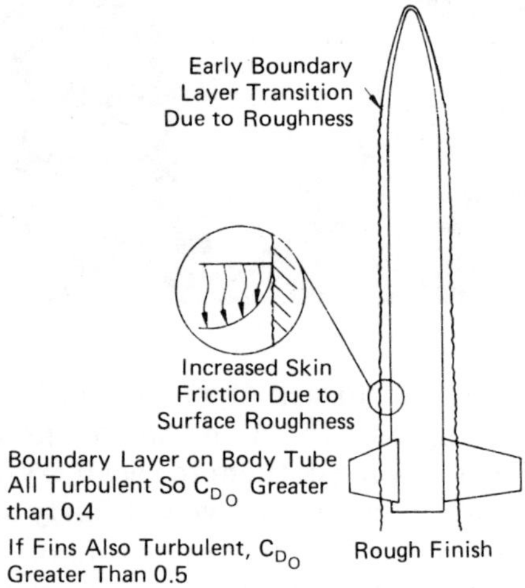

Figure 6-13. Laminar flow changing to turbulent flow.

The air acts as if it were in layers ("laminae"). The molecules in each "layer" slide over the other molecules. The molecules in the "layer" next to the surface essentially have 0 velocity, and each succeeding "layer" further from the surface has a higher velocity of motion relative to the surface.

In turbulent flow, the motion of the molecules is much less regular because of the mixing of the different layers and irregularities in the velocity of the molecules at different distances from the surface. This turbulent flow does not have the smooth variation in the layers' speed of flow as the distance from the surface increases, but instead it has a sharp decrease in velocity of flow very close to the surface and much mixing (turbulence) of the molecules at different distances from the surface. Essentially, laminar flow is a smooth flow, and turbulent flow is a flow characterized by an uneven mixing of layers (see Figure 6–13).

PRESSURE DRAG

★ The force resulting from an object "pushing" its way through a fluid is pressure drag.

The pressures on a moving object vary with the the object's direction of motion and its size and shape. The pressure drag is the "pressure" that retards the motion of the object.

FRICTION DRAG

★ The retarding force produced by the object sliding past the molecules of the fluid it is moving through is friction drag.

The amount of friction drag produced by the motion depends upon the amount of surface exposed to the motion of the fluid (the "wetted surface"), the density of the fluid, and the viscosity of the fluid.

VISCOSITY

★ Viscosity measures the resistance to motion of a fluid moving over a surface.

Low-viscosity substances flow easily. Air has very low viscosity. Water has low viscosity. Substances that flow or "pour" less readily, such as thick motor oil, have higher viscosities. The lower the viscosity of the fluid an object is moving through is, the less the drag created by the motion.

EFFECTS OF SURFACE SMOOTHNESS

★ Fluids tend to retain a laminar flow pattern longer over a smooth surface than over a rough surface.

The smoother the surface, the greater the distance is along the surface over which the fluid flow will remain laminar. Generally, the smoother the surface, the less is friction drag created by the fluid moving over the surface. The shape of the surface, however, has an effect upon the drag generated by the moving object (see Figure 6–14). The longer the flow can be kept "attached" to a surface, the greater is the distance over which the object will experience laminar flow. Once the flow becomes detached or separated from the moving object, the flow becomes turbulent.

Producing an area of greatly lowered pressure "behind" a moving object causes a pressure imbalance on the object and increases the drag experienced by the object.

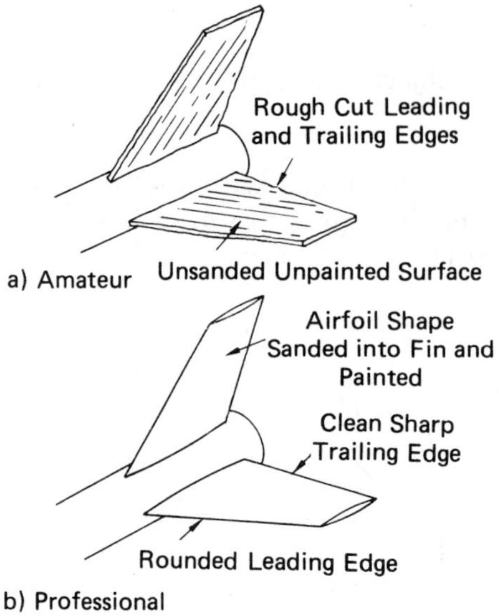

Figure 6-14.
Fin workmanship.

EFFECTS OF VELOCITY CHANGES

★ The higher the velocity, the greater is the drag produced.

The basic drag formula allows for the effect of velocity on drag:

$D = C_D A \frac{1}{2} \rho v^2$

- D — the drag force
- C_D — a dimensionless "aerodynamic drag coefficient" determined by the shape and surface smoothness of an object
- A — the cross-sectional area of the body tube for model rockets
- ρ — the density of the medium the object is moving through; pronounced "row" from the Greek letter rho
- v — the velocity of the object in motion

The drag experienced by the object directly varies with the square of the velocity of the moving object. This means that if the velocity of the object doubles, the drag of the object increases by a factor of four. If the velocity of the object triples, the drag of the object increases by a factor of nine (three squared).

ALTITUDE AND TEMPERATURE CHANGES

★ As the density of the air increases, the drag experienced by the object increases.

The higher above sea level (higher altitude), the less is the density of the air, so the drag experienced by the object is also less.

The warmer the air is in an area, the less is its density, so the less is the drag experienced by the object moving through the air.

To maximize the effect of the total impulse available from a model rocket engine, the rocket should be launched from as high an altitude as possible and in air as warm as possible.

BALLISTIC COEFFICIENT

★ Model rockets with high ballistic coefficients will reach higher altitudes than model rockets with low ballistic coefficients, when they weigh the same and are launched with the same type of engine.

The ballistic coefficient of a model rocket is the ratio of the rocket weight divided by the drag form factor. The drag form factor is determined by multiplying the drag coefficient by the frontal area (C_DA). The ballistic coefficient is important in determining the best shape and weight for a model rocket.

THRUSTING PHASE AND COASTING PHASE

★ Since the mass of a model rocket changes as its propellant is burned and expended, the ballistic coefficient of a model rocket also changes during flight.

The average weight of the rocket is used in calculations for the thrusting phase of flight. The ballistic coefficient during thrusting is determined by this formula:

$$\text{Thrusting} = \frac{W_I - \frac{1}{2} W_P}{C_D A}$$

W_I the initial weight of the model rocket

W_P the weight of the propellant in the engine

This formula adjusts the thrusting ballistic coefficient to use the average weight of the model rocket (its weight when half the propellant is gone). The solution of this equation produces an answer usually expressed in ounces per square inch. This provides a better understanding of the mass per square inch of frontal area. The higher the thrusting ballistic coefficient is, the faster the rocket can travel for a given amount of thrust.

$$\text{Coasting} = \frac{W_I - W_P}{C_D A}$$

The formula for the coasting ballistic coefficient uses the total weight of the rocket after all the propellant is gone. The higher the coasting ballistic coefficient (the greater the mass per square inch of frontal area) is, the greater the distance the rocket can coast for a given velocity at burnout. The higher the coasting ballistic coefficient is, the lower the drag.

For a much more detailed examination of aerodynamic drag and its effects on model rocket flight, read Technical Report TR-10 *Altitude Prediction Charts* and Technical Report TR-11 *Aerodynamic Drag of Model Rockets,* published by Estes Industries.

BERNOULLI'S PRINCIPLE

★ When the velocity of a fluid is increased, the lateral pressure exerted by the fluid is decreased.

Bernoulli's principle, or theorem, expresses relationships between the velocity and pressure of a smoothly flowing (laminar flow) fluid. For a horizontally flowing fluid, the total mechanical energy possessed by the fluid (the energy associated with fluid pressure, the gravitational potential energy of elevation, and the kinetic energy of fluid motion) remains constant. If the speed of horizontal flow increases, the lateral pressure exerted by the fluid decreases. If the speed of horizontal flow decreases, the lateral pressure exerted by the fluid increases. (See Figure 6–15.)

For a fluid flowing through a horizontal pipe of varying cross-sectional area, an increase in the speed of flow caused by a section of smaller diameter will result in a decrease in the pressure the fluid exerts against the inside surface of the pipe. This effect is sometimes called the Venturi effect.

The change in velocity of the air moving past an object caused by the object's shape produces a change in the lateral pressure exerted on that object by the air. This characteristic of moving air enables airplane wings to produce lift and many other features.

The changes in rate of flow of the hot exhaust gases produced by the combustion of the fuel and oxidizer in a rocket engine as they go through the nozzle are explained by Bernoulli's Principle (see Figure 6–16).

Figure 6-15. Bernoulli's principle.

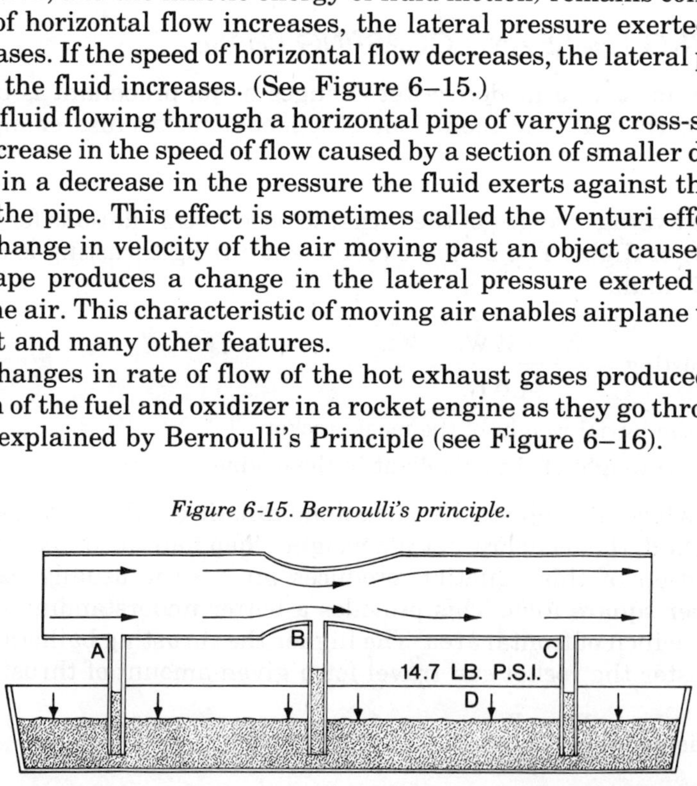

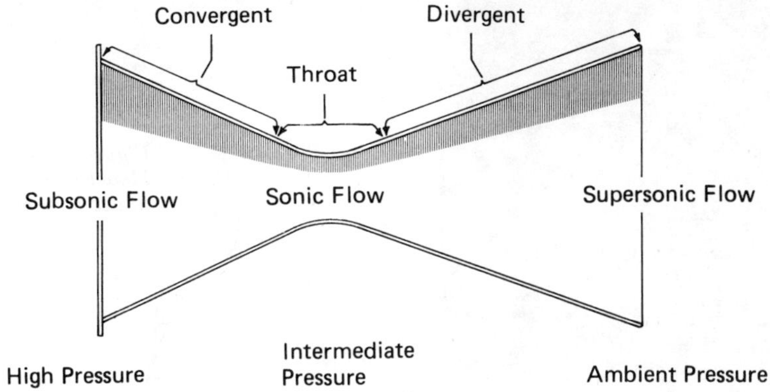

Figure 6-16. De Laval nozzle.

RECOVERY SYSTEMS

★ The recovery system slows the descent of the model rocket to permit safe recovery.

PARACHUTE

★ A parachute is a device that produces great drag to slow the descent of the model rocket.

Usually stowed inside the body tube during the thrust and coast phases of the flight, the parachute is pushed out of the body tube by the ejection charge. The parachute opens to produce a very large surface at right angles to the rocket's descent. This large, unstreamlined surface is usually in the shape of a hemisphere when fully opened or deployed.

Most parachutes used in model rockets are made of very light plastic. The typical arrangement is for the parachute to be hexagonal in shape with the shroud lines attached to each corner of the parachute. Other shapes, such as square or octagonal, are possible, but the hexagonal shape is most popular (see Figure 6–17).

Between the parachute packed in the body tube and the engine is a layer of flameproofed recovery wadding. The thickness of the layer of recovery wadding will vary, but it should usually be about one and one-half times the diameter of the body tube. It is very important to put the correct amount of recovery wadding into the body tube before inserting the parachute.

The recovery wadding should be loosely packed. If the wadding is packed too tightly, the wadding may act as a cork preventing the ejection charge from pushing off the nose cone and ejecting the parachute. If the wadding is packed too loosely, or if there is not enough wadding, the hot

Figure 6-17. Parachute with canopy open.

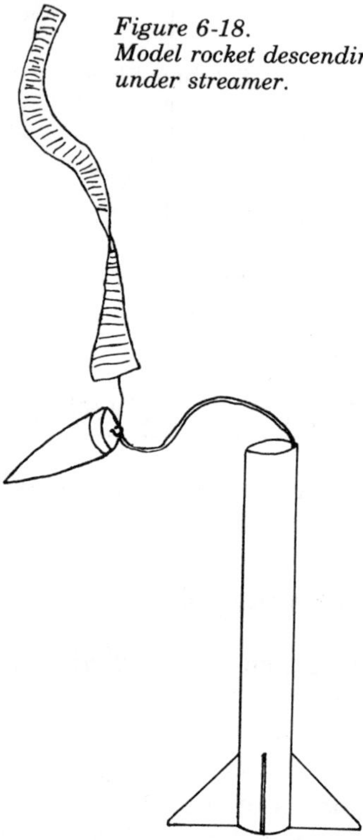

Figure 6-18. Model rocket descending under streamer.

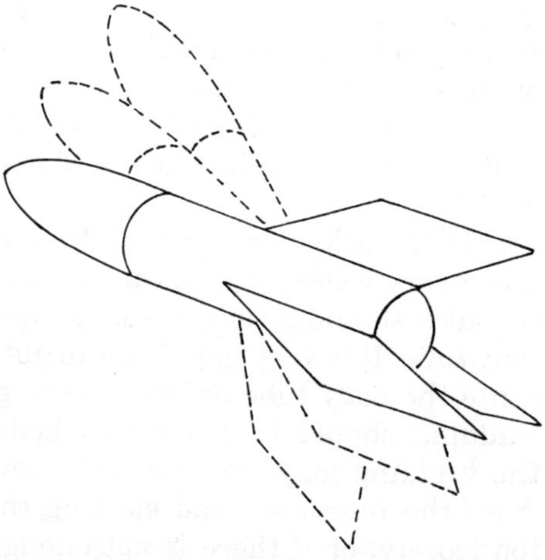

Figure 6-19. Rocket descending using tumble recovery.

gases from the ejection charge can pass through the wadding layer and either burn tiny holes in the parachute or melt the parachute into a lump. None of these conditions are desirable.

Parachutes can be made from nylon or other suitable materials. Such parachutes are often more durable than plastic 'chutes, but they are usually bulkier and heavier.

Pack the parachute very carefully *each* and *every* time. The survival of your rocket depends on that!

STREAMER

★ The effectiveness of the streamer in slowing the descent of the model rocket is chiefly determined by the streamer's surface area and its roughness

A streamer does not have the high drag properties of a parachute. Streamers are very useful for returning model rockets safely with minimum lateral drift. On windy days, a model rocket can drift sideways a long way during its descent. Using a streamer instead of a parachute permits the model to return safely yet fast enough to minimize lateral drift. The size of the streamer needed primarily depends upon the weight of the object being returned. It is quite practical to return the body of the rocket beneath one streamer the nose cone-payload under a separate streamer (see Figure 6–18).

If your parachutes and streamers are each attached to a snap swivel, you can easily change the parachute on the model for a parachute of a different size or a streamer in seconds. You can select the best parachute or streamer for each flight with a specific engine and with specific wind conditions by attaching the proper streamer or parachute just before flight.

Streamers are usually packed by rolling them into a compact roll or into two compact rolls. Lubrication of the streamer with talcum powder is rarely done, but it can be helpful.

TUMBLE

★ Model rockets that use the tumble recovery system are unstable during the recovery phase of the flight.

Tumble recovery rockets tumble as they descend. The center of gravity is not ahead of the center of pressure, so the aerodynamic forces in operation during recovery do *not* realign the rocket as it falls so that the nose of the rocket precedes the tail (see Figure 6–19).

Since the model rocket that uses tumble recovery is unstable during the recovery phase of the flight, the rocket cannot "streamline" in, but instead it flops around in the sky. This erratic motion produces very high drag and slows the rocket's descent. This method of recovery is rarely used on models that are not simple in design and relatively sturdy in construction because the rate of descent is usually higher than it would be for the same model descending under a parachute or a streamer.

BREAK-APART

★ Model rockets that use this recovery system separate into two parts at apogee.

The two parts remain attached by a shock cord. The parts are unstable, so they descend as two parts returning by tumble recovery (see Figure 6–20).

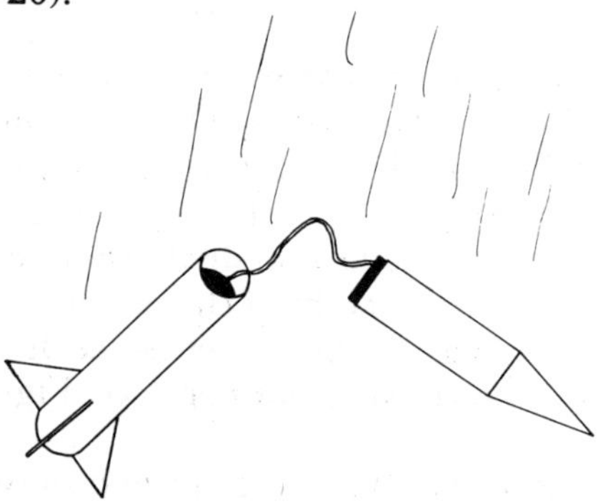

Figure 6-20.
Model rocket returning via break-apart recovery.

FEATHERWEIGHT

★ Model rockets that use featherweight recovery are so light that their descent is relatively slow.

They have adequate frontal surface area and minimal weight so that their descent, even though stable, does not reach speeds high enough to cause damage on impact with the ground. Only extremely light models can use this recovery system. The engine is ejected when the ejection charge functions. The models have to be well built, even though they are very light, to survive this type of recovery (see Figure 6–21).

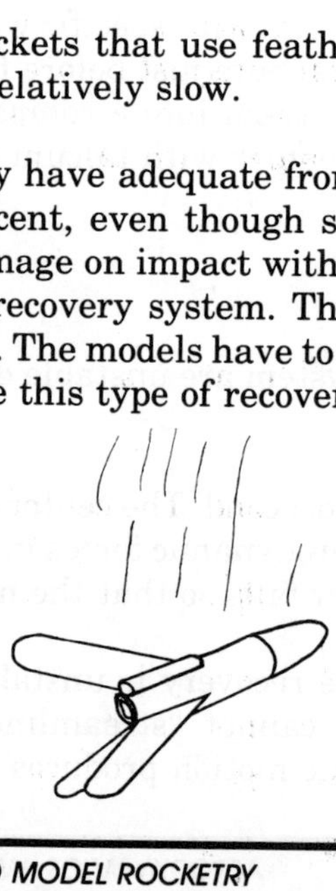

Figure 6-21.
Model rocket returning by feather-weight recovery.

Model rockets recovered by featherweight recovery must have a very low ballistic coefficient during recovery (low mass to frontal area ratio).

HELICOPTER

★ Model rockets recovered by helicopter recovery systems do not reach a high speed during descent because their configuration produces spin during descent.

This rotation produces high drag and prevents the models from falling rapidly enough to suffer damage on impact. In addition, it's fun to watch.

Most helicopter recovery models either deploy rotating vanes at apogee or by some means cause fins to go from a streamlined, low-drag design to a design with a high lift in one direction (relative to the axis of the rocket), producing a high rate of spin during descent (see Figure 6–22).

GLIDE

★ Model rockets that use the glide recovery system move forward as they descend. This horizontal motion decreases the rate of fall.

There are a number of different designs that produce this gliding motion. Their common characteristic is that the models move forward during recovery like an airplane coming in for a landing (see Figure 6–23).

Most model rockets that use the glide recovery system during descent are shaped much like airplanes. They move forward along their longitudinal axis as they descend. They also sink (due to the force of gravity) as they move forward. The glide ratio is the horizontal distance an object will move

Figure 6-22. Gyroc during boost and during descent.

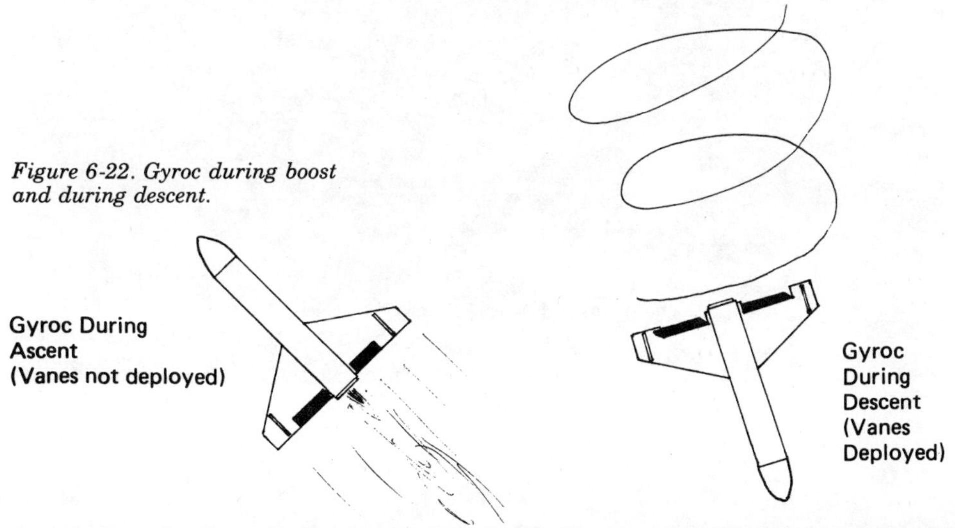

Figure 6-23. Glider being placed on launch pad.

Figure 6-24. Glider descending.

forward in still air for a given amount of altitude lost (vertical distance). The rate at which they sink or descend is called the sink rate. The lower this sink rate is, the longer the time they can stay in the air after apogee. A model with a good glide (a low sink rate) will move far during recovery, so it may return to the ground at some distance from the launch pad. The model usually glides into the wind during recovery. Position your launch pad and launch rod properly to the wind and the recovery area to take advantage of this. It is very distressing to lose a well-built glider because it glided out of sight.

If a glider can move forward 30 feet for each 1-foot loss of altitude, it has a 30-to-1 glide ratio. If this glider is gliding at a rate of 30 miles per hour, it has a forward speed of 44 feet per second and a sink ratio of 1.5 feet per second. If it is 300 feet up at apogee and passes smoothly from launch to glide mode, it should fly 204 seconds in still air before landing (see Figure 6–24).

The "Lift" section earlier in this chapter explained how to improve glide.

Sometimes a glider will "catch a thermal." This means that the glider moves into an area of heated, rising air. The rising air moves the model upward faster than the model is descending due to gravity. Catching a thermal is great, especially if you are flying in a contest, because it greatly increases your flight time. If your model is lost from sight before it escapes the thermal, you have lost both your model and the contest, because most glide contests require that the model glider be returned to the judges after the flight.

BOOST GLIDERS

★ Boost gliders are model rockets that are launched vertically but are recovered by gliding.

Boost gliders usually detach from the propulsion unit at apogee. The boost unit usually returns under a streamer or a parachute while the glider portion glides back. Rocket gliders must retain the propulsion unit during their return so that nothing is ejected or detached.

Front-engine boost gliders (see Figure 6–25) have the model rocket engine mounted at the front of the vehicle so that the vehicle is, in effect, pulled forward during the thrust phase.

Figure 6-25. Front-engine boost glider.

Rear-engine boost gliders (see Figure 6–26) have the model rocket engine mounted at the rear of the vehicle so that the entire vehicle is pushed forward during the thrust phase.

A pop-pod boost glider (see Figure 6–27) has the model rocket engine mounted in a detachable pod that separates at apogee. The engine and the pod return to the ground through some type of recovery device. The glider returns to the ground on its own.

Another type of glider is the variable geometry glider (see Figure 6–28). In this type of design, the rocket is launched vertically, but its configuration alters at apogee to deploy gliding surfaces.

DESIGN CRITERIA FOR BOOST GLIDERS

★ An efficient boost glider must perform well as a model rocket in vertical flight. It must also be an efficient glider during the recovery phase of the flight.

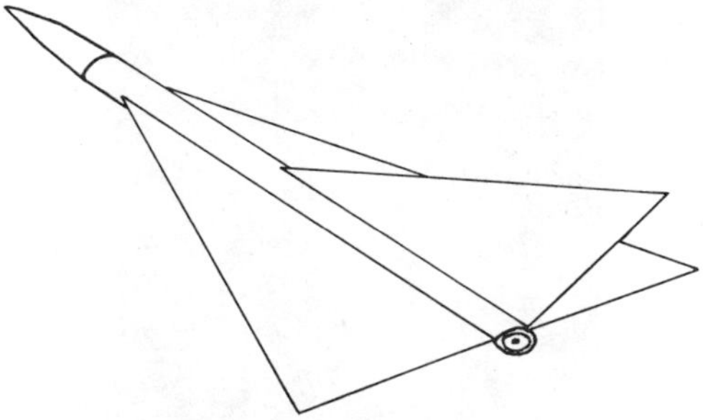

Figure 6-26. Rear-engine boost glider.

Figure 6-27. Pop-pod boost glider.

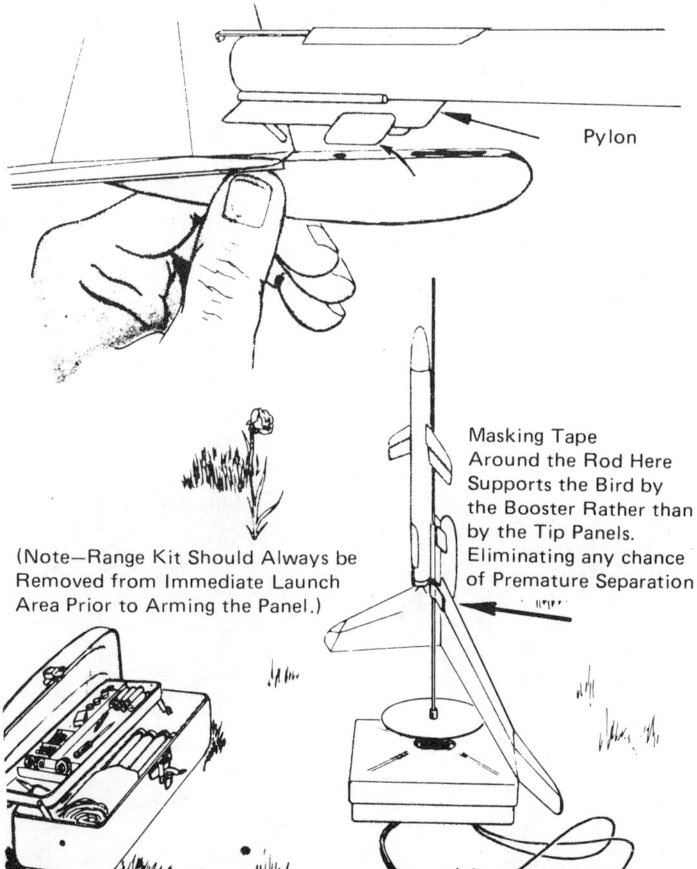

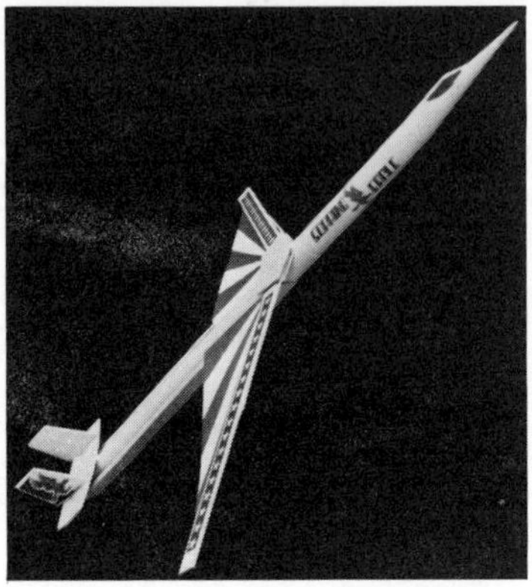

Figure 6-28. Variable geometry glider.

Figure 6-29. Airplane in flight.

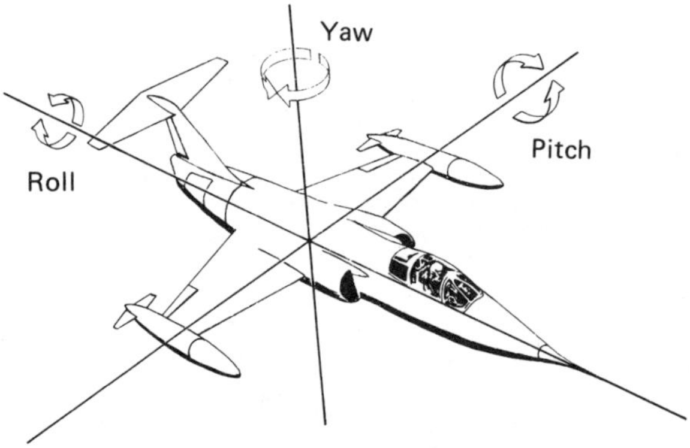

Aileron up — pressure deflects
wing downward

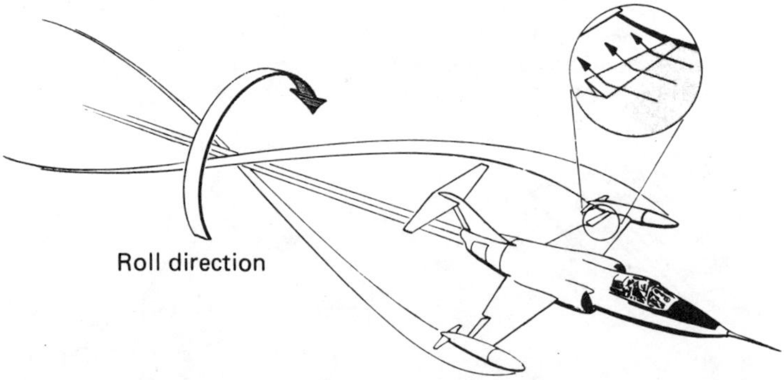

Figure 6-30. Ailerons used to produce roll.

Perhaps the best way to think of boost gliders is to regard them as airplanes and to design them for successful operation as aerodynamic gliding bodies, while making them successful model rockets.

The motions of an airplane can be explained in relation to three axes of motion. (See Figure 6–29.)

The *longitudinal axis* of motion extends through the center of the aircraft from front to rear. The vehicle may rotate or *roll* about the longitudinal axis. This movement is normally accomplished by the movement of the ailerons near the outer tips of the wings. When the ailerons are moved in opposite directions, the aerodynamic lift forces produced cause the aircraft to roll about the longitudinal axis (see Figure 6–30).

The *lateral axis* extends through the aircraft's center of gravity from side to side. Motion about this axis is called *pitch*. Pitch is usually produced by movement of the elevators on the rear of the horizontal tail surface (see Figure 6–31).

The *vertical axis* extends through the center of gravity of the aircraft from top to bottom. Motion about this axis is called *yaw* (see Figure 6–32). Yaw is produced by movement of the rudder at the rear of the vertical stabilizer at the aircraft's tail.

For stable vertical flight, the model rocket's center of gravity must be located ahead of the center of pressure. The commonly accepted criterion for

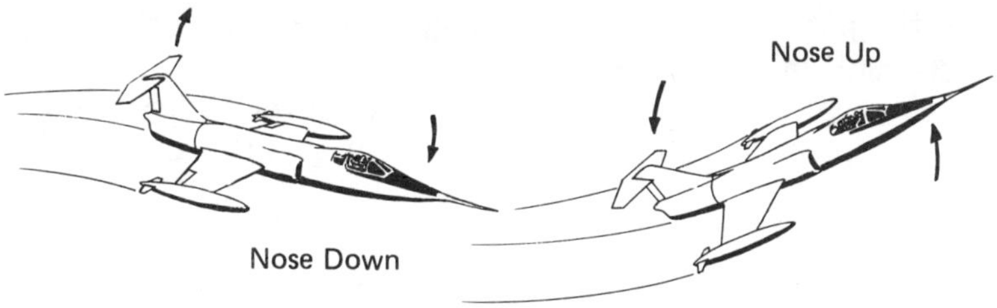

Figure 6-31. Pitch produced by movement of elevators.

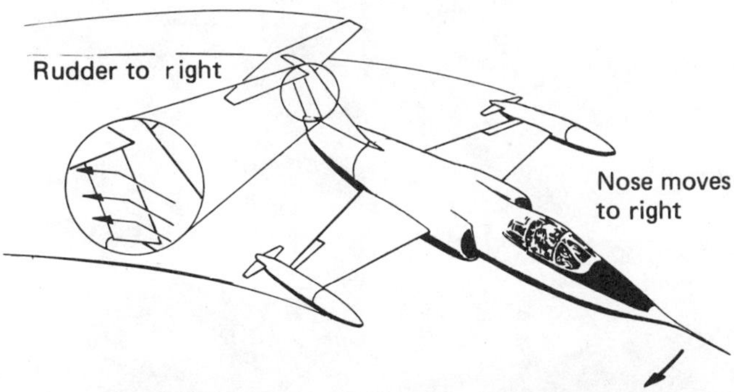

Figure 6-32. Yaw produced by movement of rudder.

stability is that the center of gravity must be located at least half of one body diameter distance in front of the center of pressure.

For gliding horizontal flight, the center of gravity and the center of pressure must have somewhat the same relationship. The center of gravity must be in front of the center of pressure as the rocket moves "forward" (in the direction of the model rocket's nose) during glide. The amount of lift produced by the wings must be very large if the vehicle is to remain long in the air. The sink rate is strongly affected by the amount of horizontal (in the glide position) surface area ("wings") and the speed of the glider. Generally, the greater the effective horizontal surface area, the greater the lift force produced and the lower the sink rate. However, this surface area produces drag, both during glide and the thrust phase of the flight.

The shape of the wings is also important in determining the amount of lift generated. The more efficient the wing, the greater the lift produced for a given amount of surface area. However, the lift force generated during glide is usually also present during the thrust phase. This force cannot be too great, or the model rocket will "loop" during the thrust phase of the flight and possibly crash.

The transition from vertical flight to gliding flight is critical. Most failures of gliders to glide occur because this transition from coasting vertical flight to horizontal glide is not made. This transition can be implemented by mechanical devices activated by the ejection charge or by changes in the position of the center of gravity as the engine is ejected or moved as a result of the operation of the ejection charge.

The design of successful boost gliders and rocket gliders is governed by scientific principles, but in practice it is still greatly an art form. Most

Figure 6-33. Wind tunnel test.

successful designs are the result of much experimentation. As research continues, this area of model rocket design will probably see significant changes in future years.

Construction and use of a wind tunnel (see Figure 6–33) can be very helpful in perfecting successful gliders. However, trial-and-error tests seem to be the most common method of design.

Much helpful information on the design, construction, and operation of model rocket boost gliders can be obtained from study of the literature in the field. Espceically helpful are Technical Report TR-4, *Rear Engine Boost Gliders;* Technical Report TR-7, *Front Engine Boost Gliders;* and Technical Report TR-5, *Designing a Wind Tunnel,* all published by Estes Industries.

7
MODEL ROCKET DESIGN

STABILITY

★ Stability is the single most important consideration in designing a model rocket.

Safety is, of course, the primary consideration in designing for stability. Such performance factors as acceleration and altitude can be affected by lack of stability.

A model rocket is a "real" rocket in miniature. The same principles governing the flight of a space shuttle govern the performance of a model rocket in flight, as explained in Chapters 5 and 6. Model rockets operate entirely within the atmosphere, and full-size rockets operate totally or partially within the atmosphere.

Within limits, model rockets duplicate the performance of their full-size counterparts. Because of weight and cost limitations, however, model rockets cannot have the sophisticated active guidance systems used by full-size rockets. It is important to understand that the relationship of the center of gravity (CG) and the center of pressure (CP) of a model rocket is a primary factor determining a model's stability.

Stability in a model rocket can be defined as the tendency of a rocket to travel in a straight course opposite the direction of its thrust despite rotating forces caused by outside disturbances. In other words, a stable rocket will travel in a relatively straight upward direction, without tumbling, and with minimum oscillation.

Figure 7-1. Airflow and rocket stability.

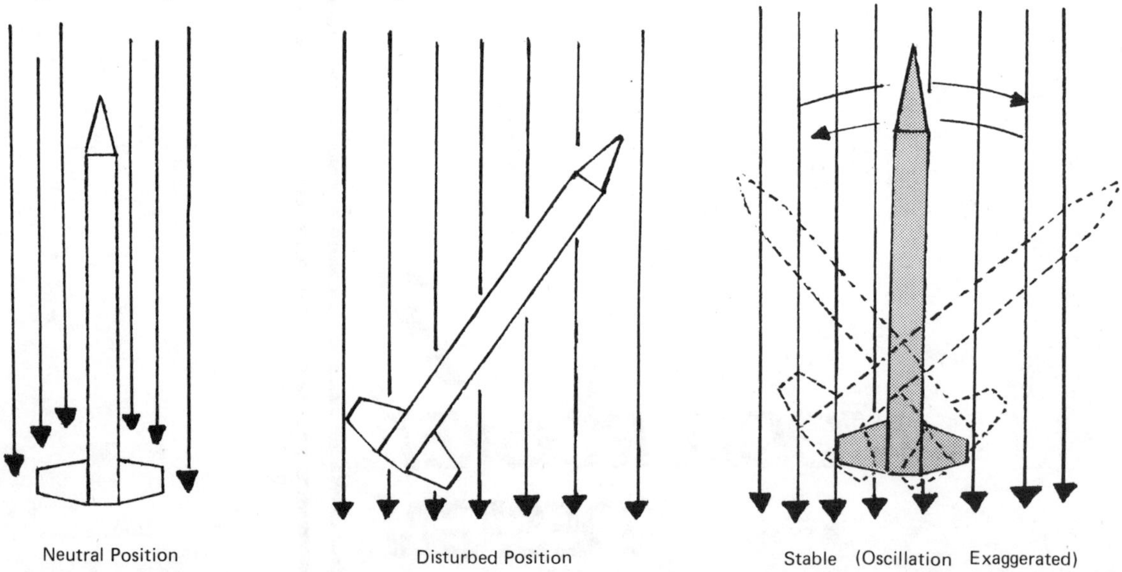

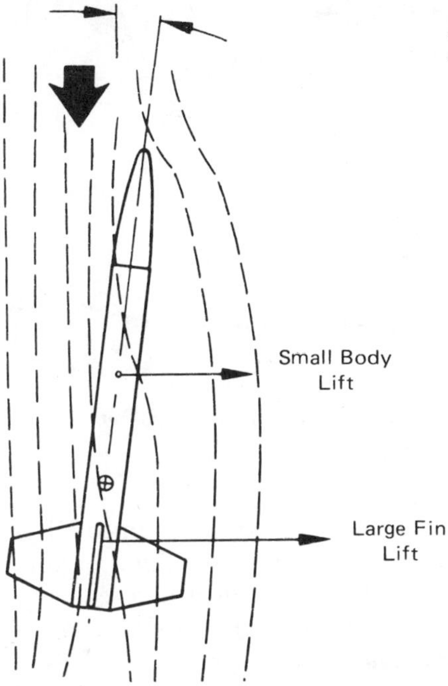

Figure 7-2.
Rocket at angle of attack.

Model rockets are stabilized primarily by the effects of air movement on their fins, which function as static control surfaces (see Figure 7–1).

The "neutral" position shows the rocket moving vertically through still air. The forces produced by the air stream moving past the rocket are equally distributed on the fins and the other components.

As illustrated here, the fins on the model rocket interact with the air through which they are moving. The aerodynamic pressures on each side of each fin are balanced as long as the rocket is moving straight. If the rocket's longitudinal axis is not parallel to the air flow, as by a gust of wind or a misaligned fin causing the rocket to rotate, the fins are no longer aligned parallel to the air flow. They are now at an angle of attack to the air stream (see Figure 7–2).

In this "disturbed" position, the angle of attack induces lift on the fin. This causes the fin to move toward the "neutral position" (fins aligned parallel to the air flow). This force causes the rocket to rotate in that direction around its center of gravity. The rocket tends to rotate past the "center position" at which the fins are aligned parallel to the air stream. The fins generate lift in the opposite direction as soon as they are past the "center position." The rocket will now rotate in the opposite direction. This process continues for several more oscillations until the oscillations are reduced and removed. This oscillation takes place at very high speeds, usually not noticeable to the eye. The more stable the rocket, the fewer oscillations before the rocket is again flying straight.

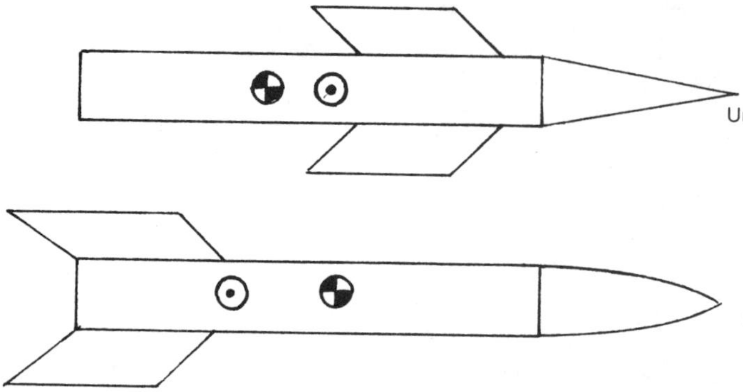

Figure 7-3. Stable rocket and unstable rocket.

In a stable model rocket, this oscillation will diminish and damp out, and the rocket will return to the neutral position. An unstable model rocket will not return to the neutral position through the forces generated by the interaction of its fins and the air stream. The unstable model rocket may oscillate in wider and wider swings until it eventually "flips over" and loses all stability (see Figure 7–3). An improper CG/CP relationship can cause this overoscillation. Such oscillation will create excess drag, retard acceleration, and reduce the altitude attained by the rocket.

A stable rocket will oscillate slightly while in flight in response to gusts of wind. The CG/CP relationship will alter slightly in flight because CG moves as the weight of the engine is reduced by the propellant undergoing combustion and being expelled (see Figure 7–4). This moves the CG forward farther from the CP. This increases the stability of the model rocket.

Once in flight, a model rocket is totally out of your control, so you must "build in" stability using a proper CG/CP relationship.

CENTER OF GRAVITY

★ The center of gravity of a model rocket is the point where a rocket's weight is equally distributed to the front and to the rear.

A model rocket will balance at its CG (see Figure 7–5). The symbol for center of gravity is ○. When checking the CG of a model, the determination should be made with the rocket fully prepped for flight, the recovery system properly packed with recovery wadding in place, and a new engine and new igniter in place.

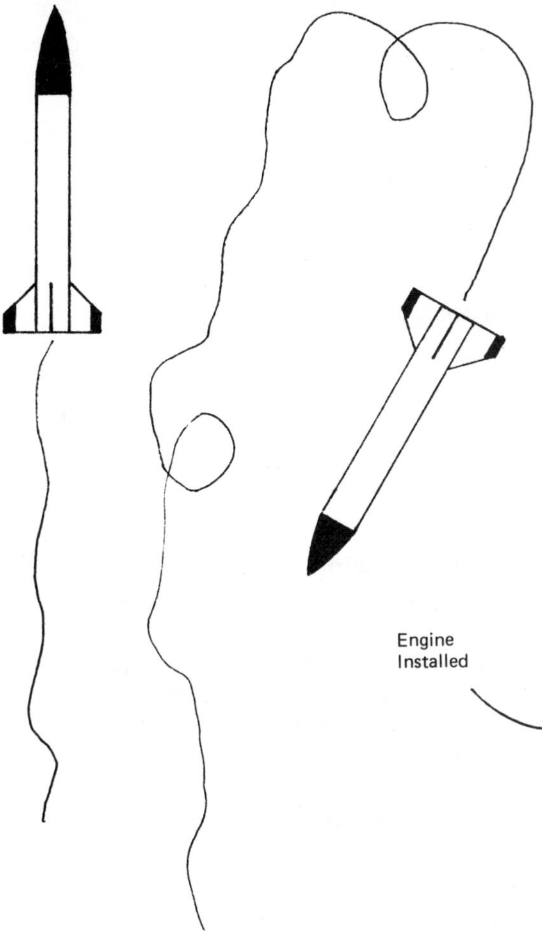

Figure 7-4 (left).
Flight path of stable rocket (far left) and flight path of unstable rocket.

Figure 7-5 (below).
Rocket being balanced at center of gravity.

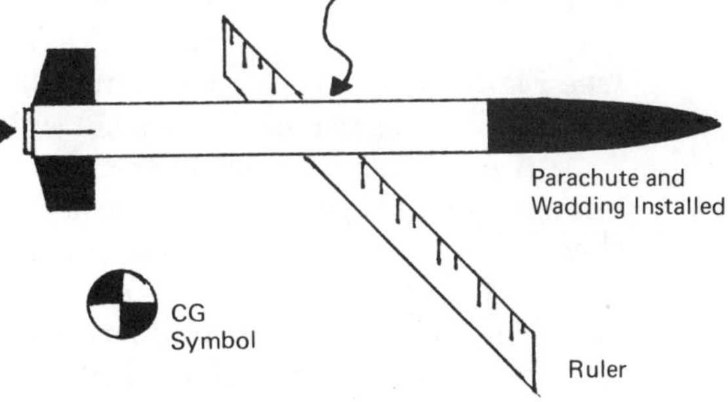

CENTER OF PRESSURE

★ The center of pressure of a model rocket is the point where all aerodynamic forces are balanced.

There is as much aerodynamic force acting on the rocket behind the CP as ahead of it. The symbol for center of pressure is ⊙.

A quick way to determine the approximate CP of a model rocket is to make an exact-size cutout of the model (a profile) in cardboard (see Figure 7–6). The point where the cutout will balance on the edge of a ruler will be the *approximate* center of pressure.

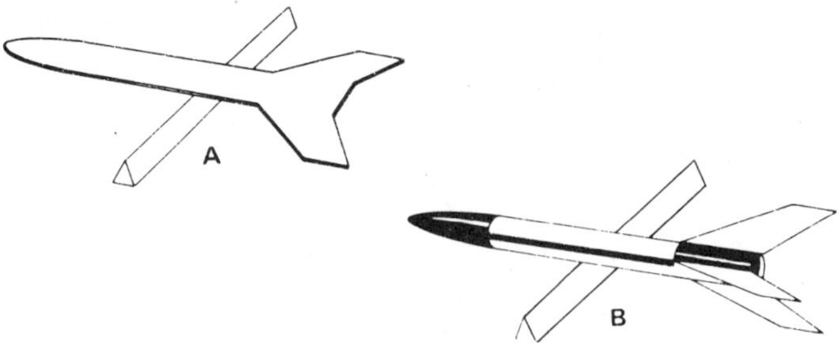

Figure 7-6. Cardboard cutout balancing on ruler.

STABLE CG/CP RELATIONSHIP

★ Ideally, the CG should be one fuselage diameter (caliber) ahead of the CP (see Figure 7–7).

A rough test of stability can be conducted by "swing-testing" a rocket. Swing-testing is accomplished by making a slip loop in a 10-foot length of string and taping the string to the rocket at the CG, then swinging the rocket horizontally with the rocket's nose pointed in the direction of flight (see Figure 7–8). If the rocket tumbles or if the base of the rocket turns into the direction of travel, the rocket will not be stable in flight.

If a rocket does not check out as stable, (meaning the CG is not far enough ahead of the CP to fly straight because the fins interact with the air flow) there are two methods of improving the CG/CP relationship: moving the CG or moving the CP.

The CG can be moved forward, if necessary, by adding weight to the nose of the rocket, usually by clay packed into a hollow nose cone or by washers added to the base of the balsa nose cone by attaching them with a screw eye (see Figure 7–9).

Moving the CP toward the rear is accomplished by moving the fins back, increasing their size, or by adding fins. The CP can be moved forward by using smaller fins. The more stable the rocket, the more it tends to rotate (weathercock) into the wind during flight.

Figure 7-7. Stable rocket showing CG/CP relationship.

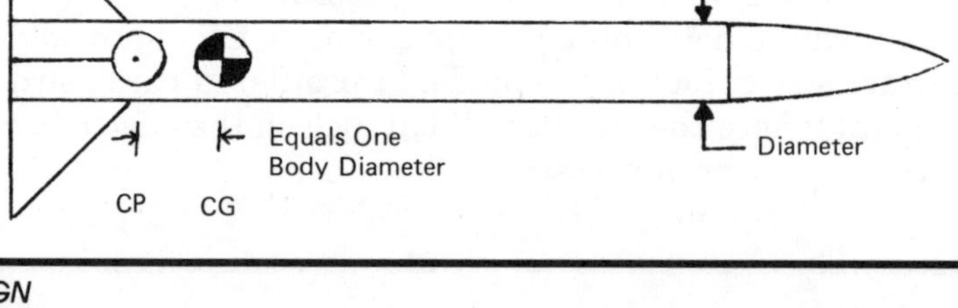

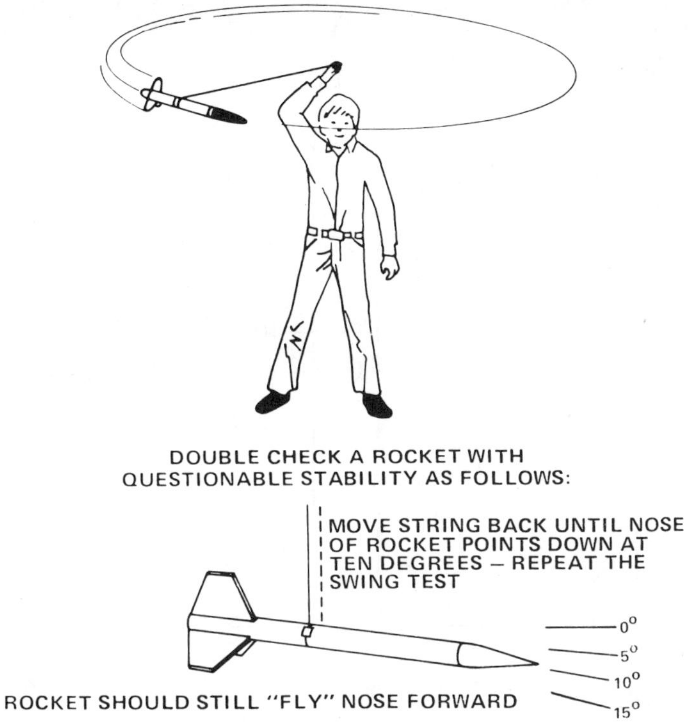

Figure 7-8. Swing test.

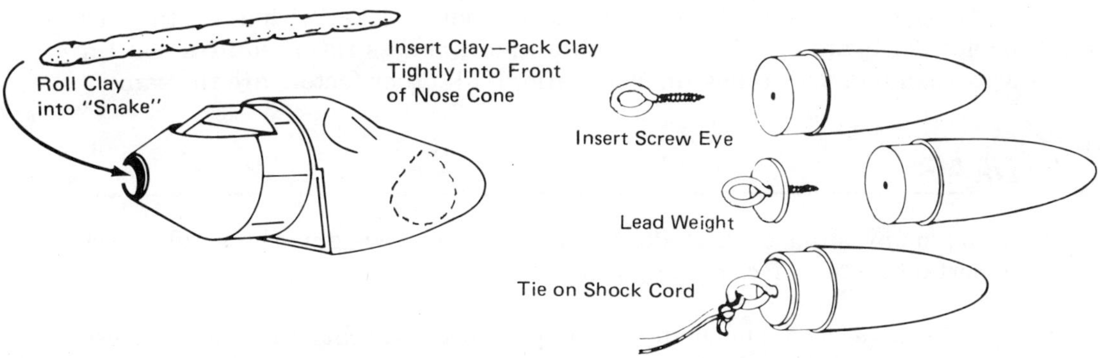

Figure 7-9. Adding clay to hollow nose cone and adding washers under screw eye.

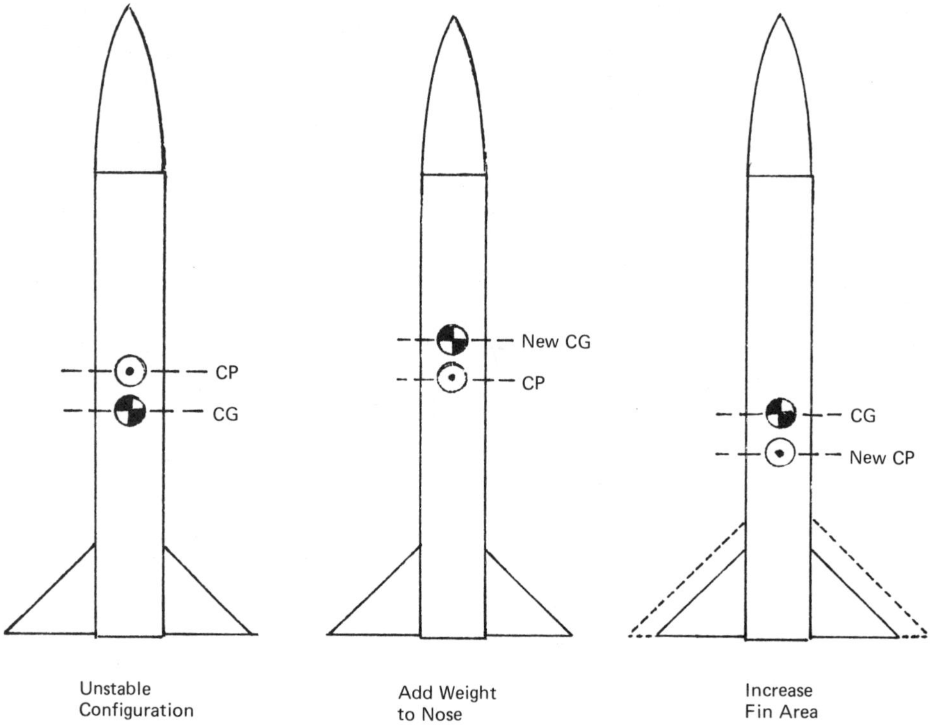

Figure 7-10. Problem CG/CP relationships and their corrections.

In general, it is best to make adjustments in the CG/CP relationship by either adding weight to the nose or by using larger fins (see Figure 7–10).

Long, slender rockets are usually more stable than short, wide rockets. Calculating the "fineness ratio," or the ratio of length to diameter in a rocket, will help you determine the potential for stability in your design. An ideal fineness ratio is 10 to 1 (a 1-inch diameter rocket with a length of 10 inches, for example). A rocket with a high fineness ratio requires less fin area than one with a low fineness ratio, if all other factors are the same.

DRAG

★ Aerodynamic drag on a model rocket is, next to weight, the single most important factor affecting performance.

As mentioned earlier in this chapter, poor stability can create excess drag. Poor design in the form of a blunt nose cone, square-edged or excessively large fins, unnecessarily great frontal area, or numerous projections from the body tube can also create excess drag.

Some drag-inducing components are an integral part of a scale or "fantasy" model, but much can be done to reduce the total drag on a typical model rocket.

TRANSITIONS

★ Smooth, unbroken flow of air over a model rocket is called "laminar" flow. Laminar flow produces minimum drag. This flow is broken up by any bumps or irregularities on the surface of the model (see Figure 7–11). Turbulent flow produces much higher drag than laminar flow.

Laminar flow can also be broken up by poor finishing, unfinished balsa, or tape. Any surface irregularities can cause the laminar flow to convert to turbulent flow.

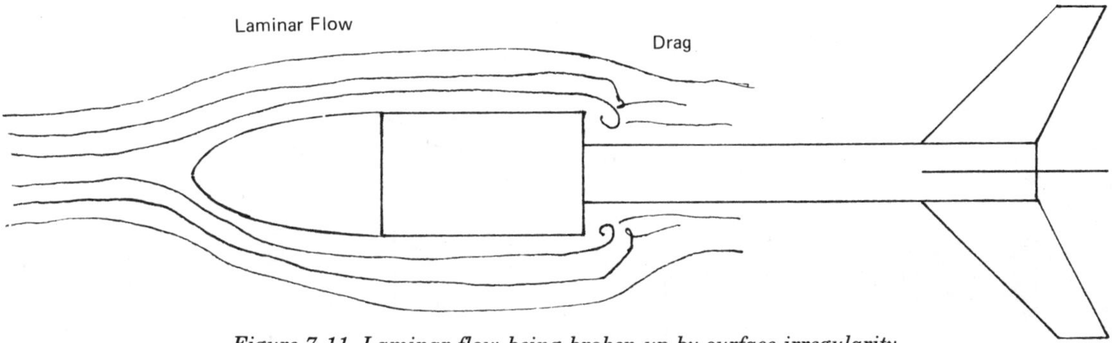

Figure 7-11. Laminar flow being broken up by surface irregularity.

NOSE CONE SHAPES

★ The shape of the nose cone has a great effect on the amount of drag that a model rocket experiences.

A parabolic nose cone creates the least amount of drag at the usual model rocket flight speeds. A square shape produces the most drag (see Figure 7–12).

A good length-to-width ratio for nose cones is three to one.

FINS

★ A sweptback fin creates less drag than a straight or forward-swept fin. Rounded corners create less drag than sharp corners (see Figure 7–13).

The leading and trailing edges of a fin have an even greater effect on drag than the shape of the fin. Square edges on a fin create the most drag,

Figure 7-12. Nose cone shapes—poor to good.

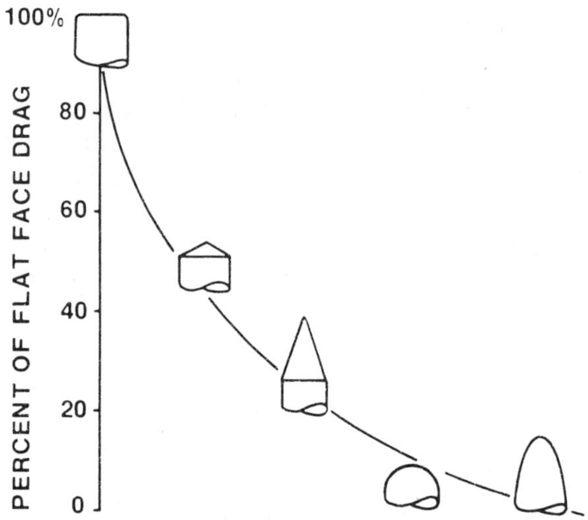

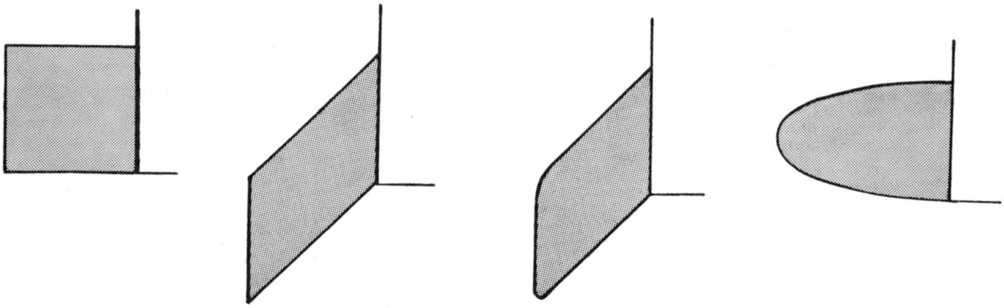

Figure 7-13. Fins—poor and good.

while rounded edges create less drag. The ideal shape for the fin cross section is a rounded leading edge and a "knife-edged" trailing edge.

The base of a fin (where the fin is attached to the rocket's fuselage) can create drag. Glue fillets at the fin/fuselage joint will help to smooth the air flow there and will reduce drag (see Figure 7–14). The fillets also strengthen the fin/fuselage joint.

Another consideration in designing fins is the fin area. For stability, the size of each fin should be at least one and a half times (in both directions) the diameter of the body. Keep in mind, too, that the span (length) of a fin is more important than the chord (front-to-back distance). The distance that the fin extends from the body has more effect on the stability of a rocket than the chord of the fin does. The farther the fin is from the center of pressure, the greater the effect of the fin due to the "lever arm" effect. The

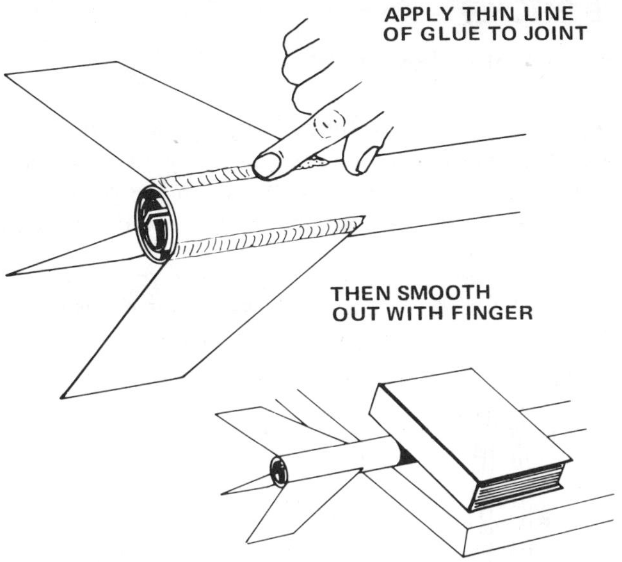

When the glue on the fin joints has dried, apply a glue reinforcement to each fin/body tube joint. Holding the model level, apply a line of glue to both sides of each fin joint and on both sides of the launch lug. Smooth out the glue with your finger. IMPORTANT -- Support rocket on table edge as shown until the glue dries.

Figure 7-14. Glue fillets.

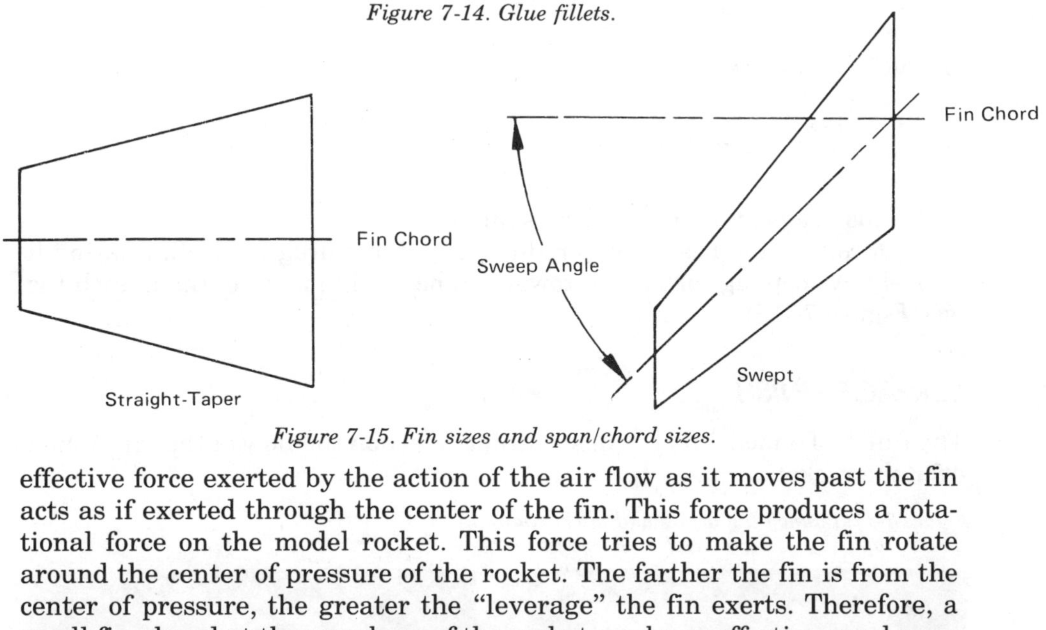

Figure 7-15. Fin sizes and span/chord sizes.

effective force exerted by the action of the air flow as it moves past the fin acts as if exerted through the center of the fin. This force produces a rotational force on the model rocket. This force tries to make the fin rotate around the center of pressure of the rocket. The farther the fin is from the center of pressure, the greater the "leverage" the fin exerts. Therefore, a small fin placed at the very base of the rocket can be as effective as a larger fin located farther forward, nearer the center of pressure (see Figure 7–15).

BASE OF THE ROCKET

★ A flat base on a model rocket will create high drag.

This is because the base produces drag just as a square trailing edge on a fin will create drag. This base drag is due to low pressure at the rear of the rocket caused by the air stream separating from the rocket. A "boat tail," or tapered base, will decrease the amount of induced drag (see Figure 7–16).

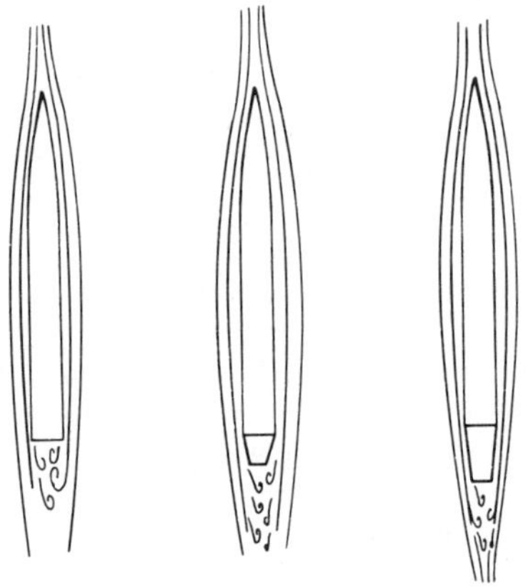

Figure 7-16.
Boat tail and no boat tail.

LAUNCH LUG

★ A launch lug can account for up to 15 percent of the total drag on a model rocket (see Figure 7–17).

This drag can be reduced by shaping the lug.

Of course, the best way to reduce launch lug drag is by eliminating it entirely. A "pop lug" or launch tower can be used in place of the launch lug (see Figure 7–18).

SURFACE FINISH

★ The finish of a model rocket has a strong effect on the flow of the air stream over the rocket.

Figure 7-17. Launch lug before and after trimming.
(Photographs by J. Kevin Wolfe, ARTifacts.)

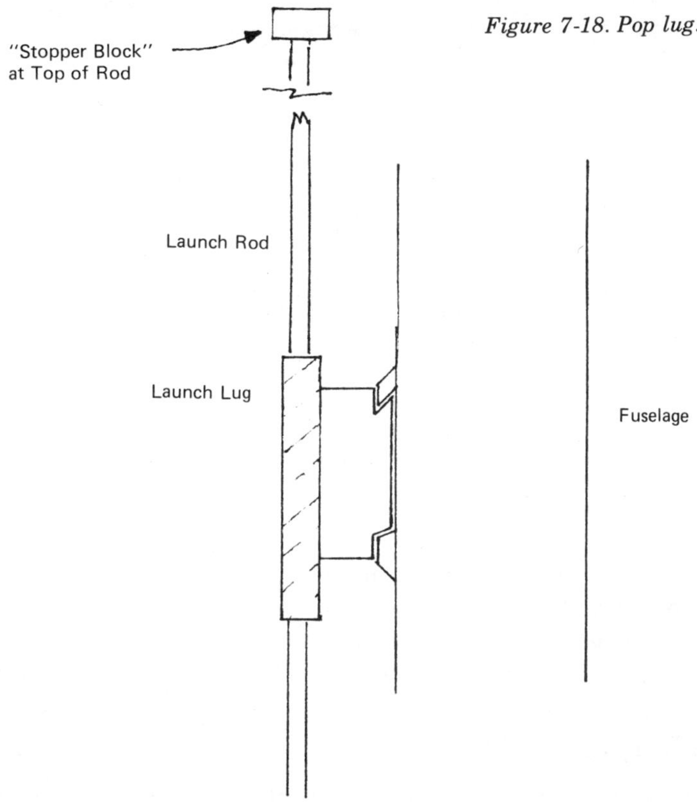

Figure 7-18. Pop lug.

Every effort should be made to keep the air flow laminar on the rocket surface as far as possible. Unfinished or poorly finished balsa parts are big culprits in creating drag. Usually, very fine sanding and several coats of balsa filler coat or sanding sealer, well sanded with fine grades of sandpaper, will eliminate drag created by rough balsa surfaces. It is possible to achieve a "plastic-smooth" balsa surface with some effort.

Cardboard or paper parts are usually prefinished and will not create much drag unless they are so thin that they warp in the air stream. If this is the case, not only will the rocket have excess drag, but it may become unstable. If you are going to use cardboard or paper components, be sure that they are thick enough not to warp or that you have reinforced them by layering or covering them with a small layer of epoxy.

Plastic parts are usually very smooth, prefinished, and provide little or no problem with drag. Check plastic components to make sure that they are not warped and have no "flash." Warping can usually be corrected by gentle, repeated pressure. Application of gentle heat can also help in changing the shape of the plastic. Trim and sand "flash" from plastic parts carefully.

Glue fillets, as mentioned earlier, are an excellent way to reduce drag because they create a transition at joints.

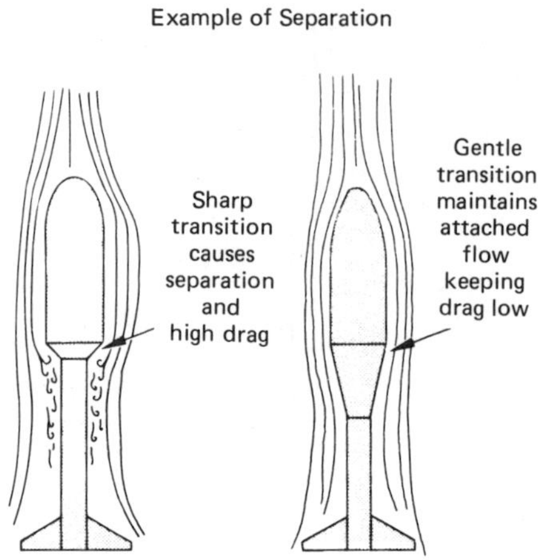

Figure 7-19. Transitions.

If the nose cone and body tube have different diameters, perhaps reshaping the base of the nose cone or making a paper shroud can help make smooth transitions to reduce drag. Transitions to make the reduction or increase in body diameters gradual can also reduce drag (see Figure 7–19).

STRENGTH AND SAFETY

You should in no way compromise strength of materials or joints in favor of reduced drag. If you cut the line too close in the amount of glue you use or in selecting weak, smooth materials over strong, rough materials (paperboard vs. balsa, for example), you are not doing yourself a favor. There is a chance that the weaker materials or construction techniques will result in the model's ripping apart during flight, and then the savings in drag won't matter.

MATHEMATICAL ANALYSIS OF DESIGNS

PREDICTING PERFORMANCE

★ Weight, drag, engine type, degree of stability, and wind are all important factors affecting model rocket performance.

Weight is a major consideration in model rocket performance and safety. It is possible to estimate the weight of a model rocket before building it by weighing samples of the materials to be used in construction and adding the weights of the samples. If you are using balsa wood for fin material, for example, you can weigh a 1-square-inch sample of the balsa you will use and then multiply this by the total area, in square inches, of the fins in the design. This will give you a close approximation of the total weight of the fins on the finished model rocket. Body tubes can be weighed, or a catalog can be used to determine the weight of a certain length of tube. The weight-per-inch value can be calculated and multiplied by the length in inches of the body tube you will use to determine the body tube weight.

Nose cones, engines, and special components can be weighed separately and the weights of the parts totaled to give the total weight of the rocket. An estimate will have to be made of the dried weight of glue. The weight of the rocket, with engine(s) and recovery wadding, can then be determined for use in predicting the altitude expected with various engines.

Measurements can be more conveniently made in metric units, such as centimeters and grams. Some model rocket company catalogs supply sizes and weights for various components used in construction. Each part can be weighed and measured if the required data is not found in a catalog.

8
SOME ADVANCED TECHNIQUES

STAGING

★ Staging permits rockets to reach very high altitudes.

Staging is accomplished by having a series of model rocket engines that ignite in sequence. The first-stage engine is ignited by standard electrical ignition. Each subsequent engine is ignited by the booster engine immediately below it. (See Figure 8–1.)

A staged model rocket must be stable in each of its flight configurations. The final stage must be stable with the heaviest engine it will ever be flown with in the engine holder. The rocket must be stable with the booster(s) in place with the largest engine(s) they will ever be flown with in the engine holder. This requires that the model have fins on the final stage and on each booster. The fins must be larger on the booster than on the final stage. The rocket must be made so that in each configuration in which it will fly the center of gravity of the combined stages (with engines in place) is in front of the center of pressure. The distance between the center of gravity (CG) and the center of pressure (CP) must be adequate to ensure stability in flight. This distance will vary with different rockets, but it will generally be equal to at least one body diameter of the rocket.

Since the engines in the booster(s) and upper stage must be physically

Figure 8-1. Scorpion two-stage model rocket.

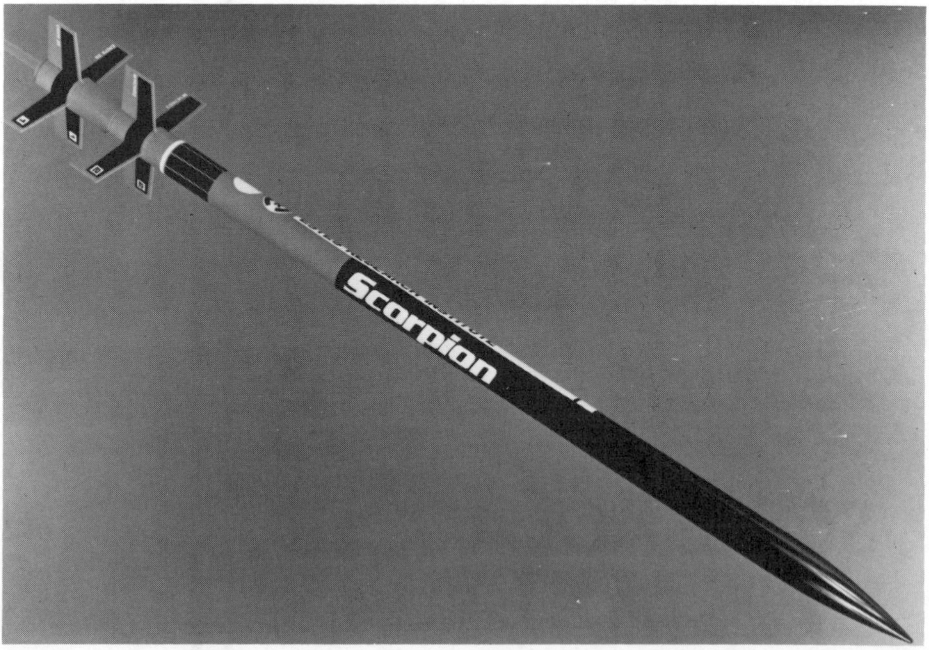

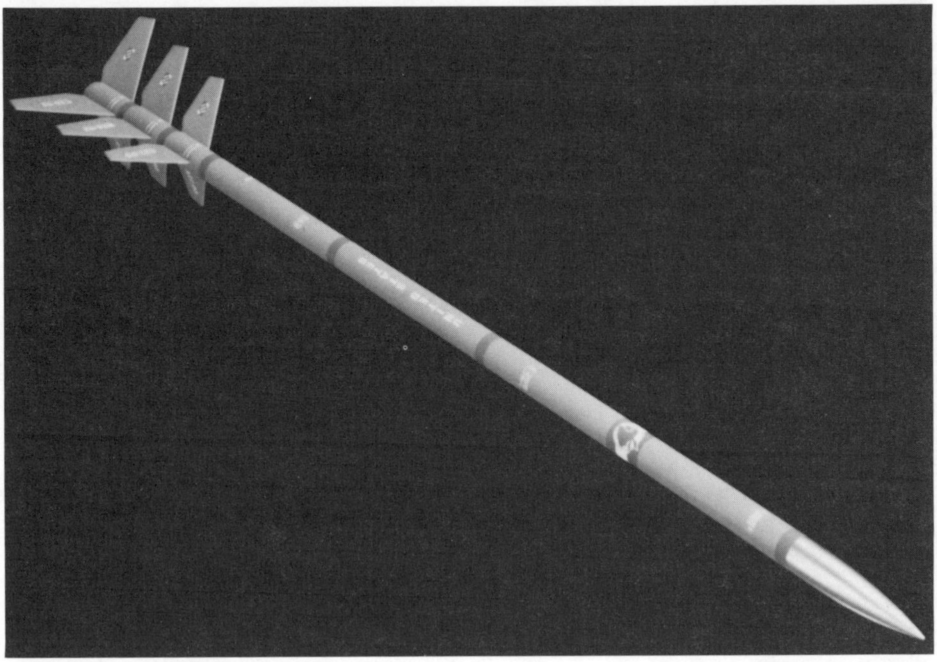

Figure 8-2. Comanche-3.

close to one another to ensure proper ignition of each stage, the weight of the rocket is concentrated near the rear. This necessitates large fins on the boosters to move the CP to the rear (see Figure 8–2).

Never fly a multistage model rocket of your own design without first making certain that it passes the string test for stability in each configuration it will fly in, with the heaviest engines that will be used in the engine holder (see Figure 8–3).

Rockets are staged to permit them to reach high altitudes. The booster engine in the first booster is responsible for launching the rocket into the air. At the moment the first booster engine uses the last of its propellant, it ignites the second-stage engine. The second-stage engine starts burning with the rocket several feet into the air and moving at a respectable velocity. The thrust of the second-stage engine adds to the velocity and altitude of the rocket. The weight and drag of the booster stage is dropped once the next stage's engine starts burning. This system permits the rocket to reach higher velocities and heights as each engine operates. It allows the rocket's final stage to reach higher altitudes than if the rocket had been launched with one engine with the total thrust available in the two or more staged engines.

The engines in each of the boosters *must* be zero-delay booster engines.

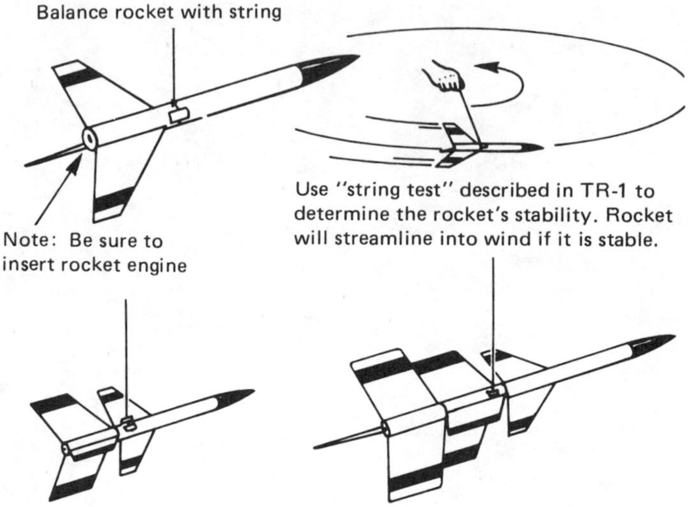

Figure 8-3. Multi-stage model rocket passing string test in each configuration.

The upper-stage engine should be an engine with a long delay. Any other pattern of engine placement will result in certain disaster!

The engines in multistage model rockets are usually secured together. This is done to increase the reliability of ignition for each subsequent stage. The zero-delay booster engine has no delay and no ejection charge. Ignition of the upper-stage engine is accomplished by tiny fragments of burning propellant from the booster engine being blown forward. This occurs as the last of the booster's propellant is consumed. (See Figure 8–4.)

As a result of the large fin area, multistage rockets are especially vulnerable to weathercocking when launched in a wind. Be very careful to launch the rockets as near vertical as possible in a wind, despite the problem of the upper stage sometimes drifting far down range during recovery.

Be careful to follow the kit instructions *exactly* when launching multistage model rockets.

CLUSTERING

★ Clusters are several engines thrusting simultaneously. The use of engine clusters permits launching of heavier payloads than could be lifted by one engine firing alone.

Using clusters of engines to launch heavier rockets or heavier payloads

MULTI-STAGING IGNITION

FIG. 1A PROPELLANT

FIG. 1B RUPTURES HERE AS BURN CONTINUES

NO DELAY OR EJECTION

The first stage of a multi-stage rocket is always ignited by standard electrical means. Second stage ignition occurs automatically upon burnout of the first stage. Figure 1A shows that the first stage engine has no delay or ejection charge. This gives instant ignition of the next stage at burnout.

FIG. 1C
WRAP OF CELLOPHANE TAPE . . .

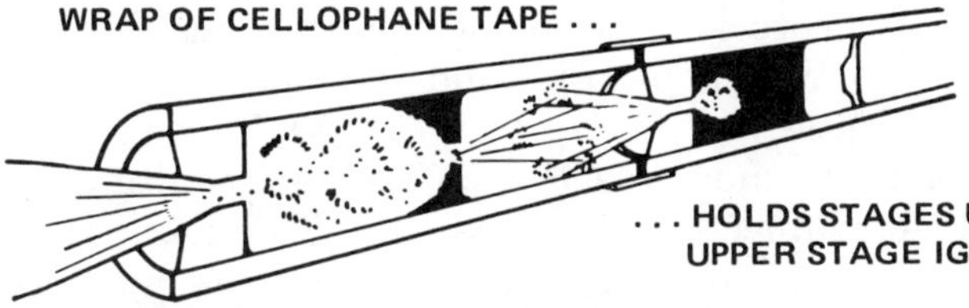

. . . HOLDS STAGES UNTIL UPPER STAGE IGNITES

In Figure 1B the propellant is partially burned, leaving a large combustion chamber. As the propellant continues to burn, the forward wall of propellant becomes thinner until it cannot withstand the high pressure inside the chamber. At this point the remaining propellant wall ruptures, and the high pressure escapes forward toward the nozzle of the next stage carrying hot gases and small pieces of burning propellant into the nozzle of the second stage engine. This action is illustrated in Figure 1C.

Figure 8-4. Stages in booster's propellant being consumed and second stage ignition occurring.

is an effective technique. Just as the development of larger, more powerful engines reduced the need for clusters of engines to launch full-scale rockets, the development of more powerful model rocket engines has reduced the need for cluster launches of model rockets.

Launching model rockets using a cluster of engines for maximum thrust is quite feasible and is very spectacular. However, it is not really needed with the D, E, and F engines available and the limitation of model rockets to 453 grams (16 ounces) total weight and a total weight of 113 grams (4 ounces) of propellant. Model rockets using the more powerful engines or engine clusters must be made very strong, especially the engine mounts and fin attachments (see Figure 8–5).

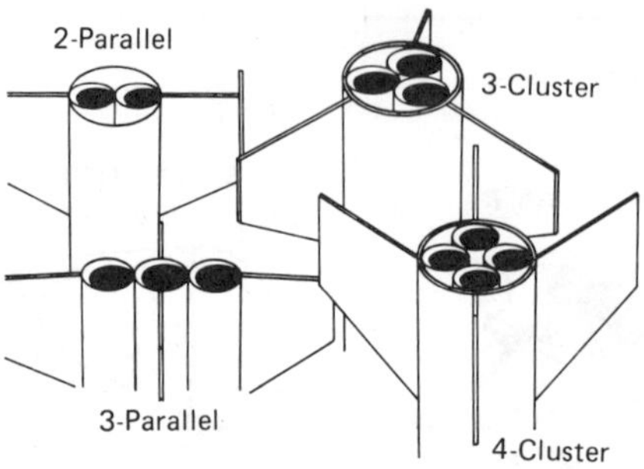

Figure 8-5. Various engine cluster configurations.

The chief problems with the use of clusters of engines for thrust is causing all of the engines to ignite at the same instant and having the engines precisely aligned so that the thrust of each engine is parallel to the longitudinal axis of the rocket. Securing simultaneous ignition requires that the igniters be wired in parallel and that gach igniter be inserted properly into each engine.

Also, the electricity must reach each igniter simultaneously and in adequate amount. All connections must be tight. This sounds simple, but it isn't, as a few model rocketeers can testify. All engines must ignite simultaneously, burn with equal thrusts, and be aligned parallel to the longitudinal axis of the rocket. If any of these conditions are not met, the rocket will experience unbalanced thrust. Unbalanced thrust can cause the rocket to move erratically, a dangerous situation. Don't use cluster power unless it is absolutely necessary.

BIOLOGICAL PAYLOADS

★ Do not launch a living animal unless it is essential to a valid scientific experiment.

Never launch a "higher animal," such as a mouse or hamster. Provide suitable protection against the g forces to which the specimen will be subjected. Make certain that the supply of oxygen is adequate for the animal in its excited state and with an adequate reserve in case the payload compartment is caught in a large tree and must be retrieved.

Rehearse the launch with the animal so that it does not become too nervous just by being handled or by being placed in the payload compartment.

Each experiment should have a "control," an identical experiment conducted in exactly the same manner but with the "variable," the thing being studied, not changed. For studying the effects of a model rocket launch on a living creature, an identical creature should undergo all of the same training, handling, and testing, except that it should not be launched. In addition to studying any changes produced in the behavior of the test specimen by being launched, you should test the control that was not launched to see if its behavior has changed.

What can you learn by launching a living animal? For the early launches in the manned space program, sometimes a chimpanzee would be launched to determine if the "hardware," the launch vehicle and the payload capsule with its life support systems, would function properly. Such tests were usually conducted just before the first launch of a human being in the capsule.

After your launch vehicle and its payload compartment are proved safe, what can you learn from the launch of a living animal? You may want to see if the launch caused a change in a "learned behavior," something the animal had been taught to do. The animal, such as an insect, may have been taught to always go the same direction in a simple two-branch maze (see Figure 8–6). When conducting such tests, you must find some means to eliminate the chance of the animal simply following an odor trail but instead always turning the same direction in the maze because that is what it has been taught.

From published reports and from correspondence received by the authors, it appears that at least 9 out of 10 launches of a living creature are successful. However, this does not mean that you should launch a mouse or hamster or similar "higher organism." Think what it would mean to you if your pet were launched—and the mission failed!

A fertile hen's egg is probably the most popular living payload to

Figure 8-6. Simple T-maze drawing. To see how fast an insect "breathes," observe how fast his abdomen (farthest part from head) expands and contracts.

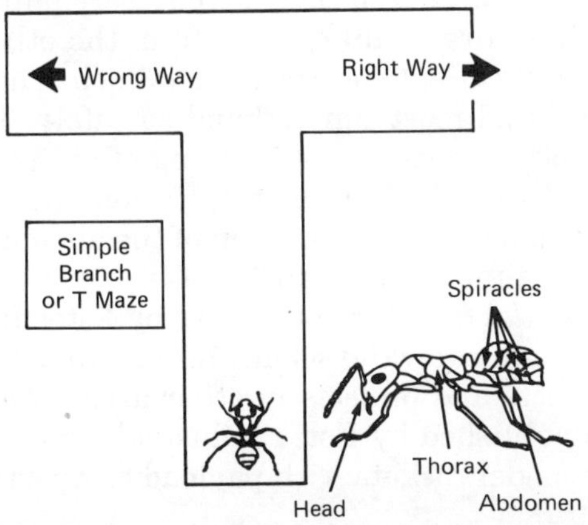

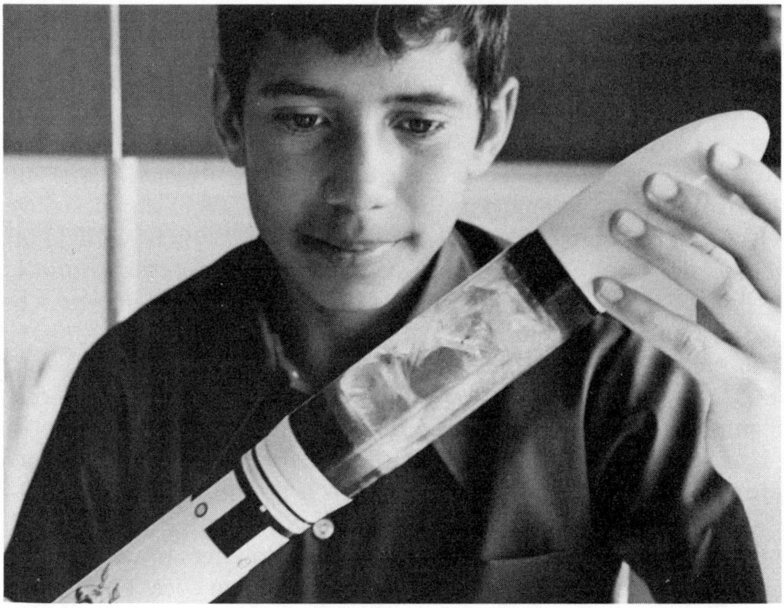

Figure 8-7. Egg in payload compartment.

launch. Secure fertile eggs from a hen or other fowl. (Smaller eggs are easier to launch in a small payload compartment.) Mark each egg to identify it, for example, with numbers put on with a grease pencil. Set aside one or more of the eggs, but launch one as the payload (see Figure 8–7). After the egg is retrieved, place it with the other eggs. Make certain that the eggs are kept warm at all times; otherwise, they will not hatch. Rotate them periodically and otherwise care for them properly. After the eggs hatch, carefully observe the chicks as they grow. Make certain that you can tell which chicks were launched. Determine if the chicks that were launched while still in the egg exhibit any measurable differences from the other chicks.

Insects make excellent specimens to launch. They are numerous, they are usually small, and most can be handled safely. Their exoskeleton provides extra protection in case your padding of the payload compartment is less than satisfactory. The rate of respiration can be easily observed by watching the expansion and contraction of the abdomen. They can, usually, be taught to run a simple two-branch maze.

Projects in Model Rocketry, published by Estes Industries, offers sound advice on how to conduct a valid scientific experiment as well as suggestions on biological projects and projects in other areas. *The Fine Art of Payload Launching,* also published by Estes Industries, should be helpful, too.

Among the model rockets with payload compartments suitable for the launch of small specimens are the Marauder (Estes #1922), the X-Ray (Estes #1218), and the Hercules (Estes #1377, used at the Alabama Space Camp).

DUCTED EJECTION

★ Ducted ejection is a system that vents the hot gases produced by the ejection charge through tubes to allow the gases to expand and cool. This technique permits safe deployment of the recovery system without the use of recovery wadding (see Figure 8–8).

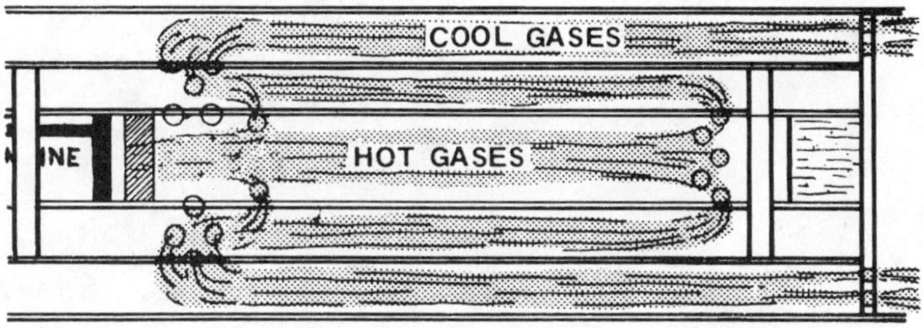

Figure 8-8. Ducted ejection.

MODEL ROCKETRY AERIAL CAMERA

Three different aerial cameras for launch as payloads on model rockets are commercially available. The first of these was the Camroc, developed by Estes Industries. This small camera was first offered in 1965. It was constructed chiefly of plastic, and took black and white pictures on circular film discs. Photos could be taken from 300 feet to 1,000 feet above the ground. (See Figures 8–9.)

In 1970 Estes Industries introduced the CineRoc aerial movie camera (see Figure 8–10). This compact movie camera used Super 8 motion-picture film. The speed at which the camera operated (31 frames per second) was higher than normal so that the film, when projected at normal Super 8 speed, produced a slow-motion effect. The sharpness of the photos and the consistency with which the camera's motor operated were sufficient to permit detailed analysis of the rocket's flight. Production of the camera ceased when the unique electric motor it used was discontinued without notice by the supplier.

The camera's lens pointed perpendicular to the camera's long axis. The lens pointed directly into a front-surface mirror mounted at a 45° angle, achieving a periscope effect that permitted the camera to photograph the ground directly below as it ascended (see Figure 8–11).

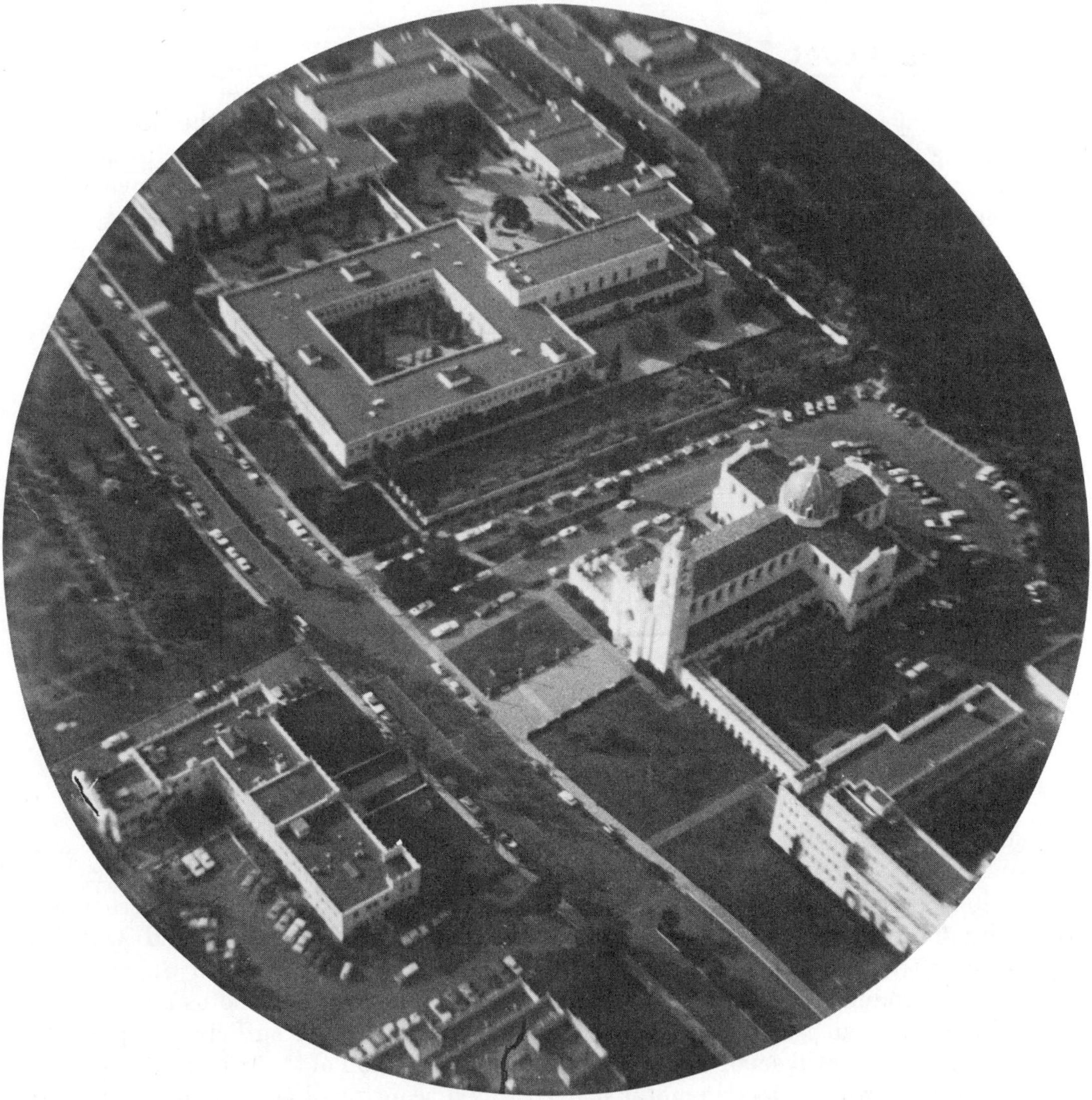

Figure 8-9. Camroc photo.

Figure 8-10. Cineroc.

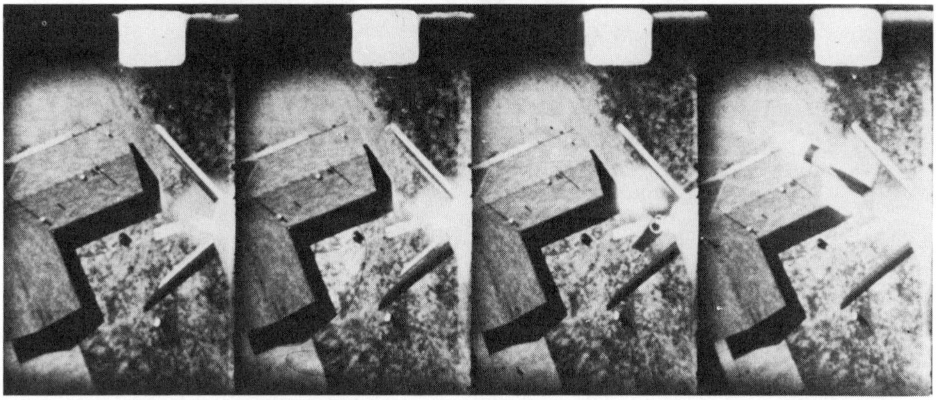

Figure 8-11. Photos from Cineroc.

Figure 8-12. Astrocam 110 on Delta II launch vehicle.

In 1979 Estes Industries introduced the AstroCam 110 (Estes #1327) (see Figure 8–12). This compact aerial camera takes pictures in full color on standard 110 film. The recommended film is Kadacolor 110 VIR 400 color negative film.

The AstroCam 110 takes excellent photos in 1/500 of a second through an acrylic lens. The camera points 90 degrees to the rocket's axis but a front-surface mirror permits a periscope effect to take photos of the area directly in front of the camera. The camera comes in kit form with the Delta II launch vehicle (also in kit form). The assembled AstroCam 110 and Delta II weigh only 3.75 ounces.

The launch vehicle that comes with the AstroCam 110 is the Delta II. The Delta II features a plastic fin unit to prevent rotation of the rocket on its longitudinal axis during flight. Rotation could cause blurring of the photograph. The usual engines for launch of the camera with the Delta II are B6-5, C6-5, or C6-7 (first flight). The engines all use a long delay because

Figure 8-13. Oblique photo taken with Astrocam 110.

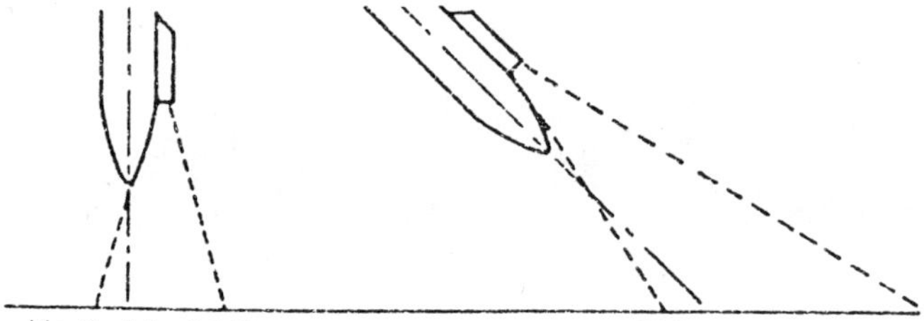

Figure 8-14. Vertical and oblique photos. Note that area photographed is not centered on centerlined camera.

the camera takes a picture forward, so the camera must be pointed downward at the moment the photo is taken if the photograph is to show a portion of Earth's surface. With a C engine, the camera will be at about 500 feet when the photo is taken.

With the shorter delay C engine (C6-5), the camera may be pointing slightly below horizontal. This produces an oblique photo of the horizon or objects on the surface but not directly beneath the camera. With the longer delay C engine (C6-7), the camera will usually be pointed downward, so the surface directly beneath the camera will be photographed (see Figures 8–13 and 8–14).

The AstroCam 110 may be launched to higher altitudes by launching it with the Challenger-II (Estes #1330). This high-performance model rocket can be flown with the D11-9 or D12-7 (first flight). Flights to 1,000 feet can be routinely made with this launch vehicle. The long delay permits the camera to tip over and point downward at the moment the shutter is tripped by the ejection charge activation. The Challenger-II uses a plastic fin unit to prevent rotation of the rocket caused by misaligned fins.

INTERPRETATION OF AERIAL PHOTOGRAPHS

★ The information derived from an aerial photograph is determined by the content of the photo and the skill of the observer.

Sometimes it is important to know how big the objects in the photograph actually are. If there is an object of known size near the center of the photo, and the photo was taken vertically, it is easy to determine the relative sizes of objects near the center of the picture by comparison. If no such "known object" is available, the measurement task becomes more complex.

The AstroCam 110 takes pictures by using a mirror, so everything on the color print is backward unless you ask that the prints be reversed when you send the film for development and printing.

The amount of information that can be derived from an aerial photograph is enormous. The obvious information includes telling just what is photographed (houses, your house, your block, your barn, a local school, and so on). In addition, the time of day can often be determined by the length of shadows. This is harder with a photo of an unfamiliar area where you do not immediately know which direction is north. (Don't forget to allow for the direction reversal from left to right.)

See Figure 8–15. From the photo below, can you tell if the area shown is a housing development, an industrial area, or "in the country"? The lack of homes indicates that it is not a residential area. Several cars are in the parking lot, so the area is at least partly commercial, a shopping center, or industrial. The area with a few weeds and grass indicates that at least part of the area is open. There are no row crops, so the area is probably not cultivated. What appear to be rows were made the previous summer to control weeds. The straight and curved "path" is made of yellow bricks and

Figure 8-15. Vertical photo of "Cape Estes," the launch pad for tours at Estes Industries.

leads to and part way around the launch pad. The launch pad is visible in the middle of the curve.

At what time of day was the picture taken? Probably near noon, since the shadows are very short.

Determination of the height from which an aerial photo was taken is possible if the picture is taken vertically and the size of one object near the center of the picture is known.

The formula for determining the height of the camera at the instant the photograph was taken is as follows:

$$H = \frac{O\,F}{I}\,.$$

H Height of camera above surface

O Object size

I Image size on negative

F Focal length of lens

Once you know the height from which the picture was taken, the formula can be modified to provide the sizes of other objects near the center of the photo.

$$O = \frac{H\,I}{F}$$

Since the height (H) is known, only the size of the image (I) under study needs to be found and the formula solved to determine the size of the object. The focal length (F) is always 30.0 millimeters.

Many details are lost when a color negative is made into a black and white print, rephotographed to proper size to fit the space available for the picture, and screened to introduce the dot pattern used in the printing process. (Look at one of these photographs with a magnifying glass. Ten power is excellent for this use.) Also, the paper on which the picture is printed is somewhat porous. Some sharpness is lost during each step in the process. Do photointerpretation from your own color prints for best results.

TELEMETRY

Model rocket telemetry to secure data from thousands of feet above the surface of the ground has been performed many times. Launching small radio transmitters was accomplished early in model rocket history. Sometimes these devices were used merely to track the rocket in flight and to help locate it on the ground after its descent. Sometimes small sensors were attached to the transmitter, which sent back information on spin rate of the rocket, air temperature, or similar measurable phenomena.

Two model rocket telemetry transmitters are the Estes Transroc and the Foxmitter. The Estes Transroc was developed for Estes Industries by Earl Estes, brother of Vern Estes, founder of Estes Industries.

The Transroc was introduced in 1971. Its 0.8 inch diameter and 4.2 inch length fit into a BT-50 body tube. It weighed 1.3 ounces with battery and antenna. It transmitted with 100 milliwatts and had a range of 1 mile. It broadcast on Channel 14 of the 27 megaHertz citizens band. In addition to its basic beacon mode (for use in finding it), it accepted attachments to detect and transmit air temperatures, spin rates, and sounds. (See Figure 8–16.)

Figure 8-16. Tranroc.

The Transroc was discontinued by Estes Industries when the Federal Communications Commission changed the frequencies available for CB use. This required modifications for the Transroc, which Estes chose not to make.

The Foxmitter was developed by Richard Fox. Its plans were first published in *Model Rocketry* in May 1979. Plans for the Foxmitter-3 were published in the September 1971 issue of *Model Rocketry*. Competition Model Rockets can supply current prices on the Foxmitter and its accessory sensors.

UNDERWATER LAUNCH

Launching a model rocket that is completely submerged in water can be a spectacular event. Setting up the proper conditions to properly produce a safe underwater launch requires quite a bit of special care (see Figure 8–17).

A large container of water is necessary. It is best not to use containers such as swimming pools because a surprising amount of undesirable residue can be left in and on the water. This residue consists of the waste products of the engine's combustion.

The rocket must be made waterproof. This can be accomplished by lightly sealing all seams through which the water might enter the rocket with a light layer of paraffin from a melted candle. The layer must not be so tight that the nose cone cannot separate easily from the body tube when the ejection charge activates. The engine and igniter must be properly placed in the engine and very carefully sealed in place. The layer of paraffin (wax)

Figure 8-17.
Underwater launch preparation.

from the candle may be much thicker here, if necessary. No water may penetrate to the propellant, or ignition will not occur. No water may penetrate the clay nozzle, or the nozzle may soften and erode away, producing off-center and/or improper thrust. Of course, all the balsa and cardboard surfaces of the model rocket must be waterproof, either through proper painting and/or through coating with a thin layer of paraffin.

Extreme care must be taken that the igniter leads and micro-clips be properly attached and not touching. If the water has minerals or ions in solution, the electricity may "short out" through the water, preventing proper ignition of the igniter.

An unexpected problem can be keeping the model rocket properly in place at the base of the launch rod. With the volume of trapped air in the body tube, the rocket may try to float like a cork! The rocket must be properly secured at the base of the launch rod so that it can travel the entire length of the launch rod, so its flight will be stable. It may be necessary to retain the rocket with a weak clothespin anchored in clay attached to the end of one fin. The strength of the retainer, whatever it is, must be just enough to prevent the rocket from floating up by itself.

One additional caution: The area around the underwater launch pad must be kept free of people for a much larger area than usual—perhaps for a 20-foot radius or more—both for safety and to prevent the spectators downwind from getting soaked! This means that you may have to use a power supply with more than the usual amperage. Perhaps two ignition systems wired in parallel will be suitable.

SCALE MODELING

Scale models appear to be exact replicas in miniature of full-sized rockets or space vehicles. The emphasis is on exact appearance rather than on optimum flight performance. Since many full-sized rockets use guidance devices not available to the model rocketeer, extra effort must be used to make certain that the scale model is stable in flight. It is frequently necessary to add weight to the nose, add accessory clear fins, or alter the scale for the fins in order to produce models with stable flight characteristics.

The range of scale models available as commercial kits is large. The exact kits available from year to year vary. What is available at any given time depends on sales trends for scale models in previous years, the popularity of specific rockets as determined by media coverage, the price points and skill levels of rockets that are currently selling well, and the success of research and development staffs at the various model rocket manufacturers in securing adequate scale data on specific rockets and in adapting this data to produce kits with reasonable skill levels and price points (see Figure 8–18). At this writing, the scale models available as kits include these kits from Estes Industries:

- Mini-Shuttle (#1391), skill level 2
- Space Shuttle Columbia (#1385), skill level 2
- Space Shuttle (#1284), skill level 4

Figure 8-18.
Estes Space Shuttle #1284.

- Cruise Missile, Boeing A. L. C. M. (#1336), skill level 4
- Sandpiper (#1389), skill level 2
- MX Missile (#1916), skill level 1

Several model rockets available as commercial kits are scaled-up or scaled-down models of other model rockets. If you have a favorite design and want to fly it in a different size or with an engine of significantly different power, this is great. Also, it is relatively easy to make scale models of any existing model rocket. Gather exact data on all sizes and shapes. Then reduce (or increase) all dimensions by the same factor. For example, to produce a three-times scale Wizard (Estes #1292), make all dimensions of the new model rocket three times the size of that dimension on the original rocket. *Always* perform a stability test on all model rockets you "scratch-build" (build from parts using your own design).

Building a great scale model depends upon successful handling of several factors. First, there must be available a good scale model kit for the rocket you want to build, or there must be available adequate scale data so you can design your own model using commercially available parts you can adequately modify for your purposes.

No matter how great the kit or the data, the construction work must be precisely done. It is fairly easy to make a great-looking semiscale model, but making a great model in accurate scale requires a tremendous amount of patience. When using a commercial model rocket kit, check the dimensions of the kit and the original rocket very carefully. Sometimes compromises are made, even on kits labeled "scale." Build everything exactly as it should be. This is especially important in the finishing of the model. Space everything precisely. Make certain that all details, colors, numbers, and so on, are correct. Modelers often choose to make a scale model of one specific rocket rather than a particular type of rocket. This degree of accuracy requires researching the markings and colors of that particular rocket as well as information on that type of rocket. Sometimes you have to do some fancy adaptations of items you can secure to make exact replicas of parts for which there are no existing parts in the correct scale.

Always test your model for stability before launching it.

Some scale competitions require that you research and then build scale models of the launch equipment used to launch that rocket. This adds much to the work required, and, for some rocketeers, to the fun.

9
MODEL ROCKET ENGINES

★ Model rocket engines are carefully manufactured miniature engines. They must meet strict standards for safety and performance.

The National Association of Rocketry establishes and enforces criteria for model rocket engines. Rule two of the NAR/HIAA Model Rocketry Safety Code states, "I will use only preloaded, factory-made N. A. R. Safety Certified model rocket engines in the manner recommended by the manufacturer. I will not change in any way nor attempt to reload these engines." Rule four of this code states, "My model rocket will weigh no more than 453 grams (16 ounces) at lift-off, and the engine will contain no more than 113 grams (4 ounces) of propellant."

CLASSIFICATION

★ Model rocket engines are classified based upon the total impulse they will produce. For convenience in design and for maximizing performance, model rocket engines are manufactured in several standard sizes (see Figure 9–1).

The code for each model rocket engine consists of three elements. The first letter or number and letter designates the total impulse (total power) produced by the engine. The second element of the engine code is the number that specifies the average thrust produced by the engine. This value is in Newtons. The third element of the engine code provides the delay time in seconds between engine burnout and operation of the ejection charge (see Figure 9–2).

Figure 9-1. Different-sized engines. (Photograph by J. Kevin Wolfe, ARTifacts.)

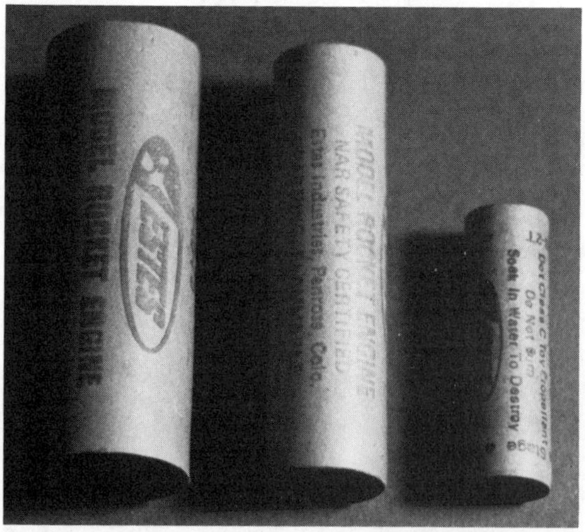

Figure 9-2. Exterior of engine showing code.

CONSTRUCTION

The solid propellant model rocket engine is a fairly complex device. The paper casing is composed of a thin sheet of paper rolled around itself with a thin film of glue keeping the layers attached to each other. This produces a strong, durable, lightweight casing. In addition, the material is relatively inexpensive, so the engines can be sold for a low price. Also, should the pressure developed inside the engine ever become too strong to be contained, the pressure can be safely released in one or more of three ways. This overpressurization is very rare, but the safety features to prevent it from causing injury are designed into the engine. The pressure can eject the ceramic nozzle from the engine, releasing the hot gases to the rear. The pressure can also eject the delay element, ejection charge, and clay retainer cap forward. This can damage the recovery system and/or the rocket, but it will not injure anyone. The pressure can cause the casing to burst. This, too, can cause damage to the rocket, but it will not injure anyone, either. The thin layers of paper making up the engine casing cannot be hurled away, causing injury to the rocketeer or a spectator.

The ceramic nozzle is designed to permit the release of the exhaust gases in a specific way. There are several different nozzle designs in use for model rocket engines. Each is based on the de Laval nozzle. In this design the exhaust gases are forced to converge as they enter the small throat of the nozzle. In the throat the gases are constricted and accelerated. As the gases leave the throat, they enter the divergent section of the nozzle before they exit the nozzle. Model rocket engines have different types of nozzles. Each type is designed to produce a specific level of thrust. By controlling the size and shape of the nozzle, the amount and content of the propellant, and the size and shape of the core, various time/thrust curves are produced.

The smoke-tracking and delay element is a propellant with one or more substances added to burn at a controlled rate. This will produce lots of smoke so that the small rocket can be located far up in the sky. After a predetermined time, depending upon the amount of smoke-tracking and delay element, the ejection charge is activated as the last of the smoke-tracking and delay element is consumed. As the ejection charge burns, it momentarily produces a relatively high level of thrust. A portion of this thrust escapes through the nozzle of the engine, but the bulk of it rushes forward under pressure, pushes off the nose cone, and ejects or otherwise activates the recovery system. The recovery wadding placed just above the engine prevents these hot gases rushing forward in the body tube from melting or otherwise damaging the parachute. Failure to insert the recovery wadding properly or to insert enough recovery wadding can allow the hot ejection charge gases to leak around or through the recovery wadding and burn small holes in the parachute, heat-fuse portions of the parachute together, or melt the parachute into a useless blob. Needless to say, none of these occurrences is desirable.

The solid propellant is composed of the fuel and the oxidizer. The fuel is the substance that burns, producing the hot gases that expand. The reaction to these hot, expanding gases exiting from the nozzle propels the rocket engine forward. The oxidizer is the chemical that supplies oxygen for the combustion reaction. In a jet engine, the liquid fuel reacts with the oxygen supplied from the air entering the jet engine. A rocket carries its own oxygen.

PROPELLANT

Important characteristics of a propellant include specific impulse, density, temperature of combustion, pressure and temperature requirements for ignition, toxicity, tolerance for handling, storage requirements, composition of combustion products, and burning rate. The burning rate is the most important of these. The rate at which the engine can produce the hot gases that produce thrust is controlled by the nature of the propellant, the area of burning surface, the pressure in the combustion chamber, the temperature of the propellant before ignition, the evenness of distribution of the particles, and the particle sizes of the fuel and the oxidizer.

Model rocket engines are very carefully manufactured. The exact chemical composition and particle size of the fuel and the exact chemical composition and particle size of the oxidizer are carefully controlled. The degree of mixing of the fuel and the oxidizer particles, and the density of the grain produced as these substances are compressed into a solid, are controlled with precision. The model rocket engine's propellant must function precisely, so all aspects of the manufacture of model rocket engines is carefully controlled. In addition to controls in every step of the manufacturing process from selection and testing of chemicals used through production of the completed engine, major model rocket engine manufacturers take three of every 100 engines produced and test them in a static thrust test stand. This device records the exact time/thrust performance of the engine as the propellant, delay and smoke-tracking element, and ejection charge operate.

If the engine does not perform within the established criteria for that type of engine, all engines made so far in that batch are destroyed, and the machine that manufactures the engines is reset. The engines produced after the resetting are carefully tested and must all meet the established criteria before any of the engines produced are saved.

One of the best ways of comparing rocket propellants is to compare their specific impulses. The specific impulse is a measurement of the propellant's efficiency. Specific impulse is determined by dividing the total impulse produced by the engine by the weight of the propellant. For example, a C6-5 engine with a total impulse of 2.25 pound-seconds and containing 0.440 ounces (0.0275 pound) of propellant has a specific impulse of 81.82 seconds. Most model rocket engines have specific impulses between 50 and 100 seconds. Estes model rocket engines have specific impulses between 80 and 83. The higher the specific impulse, the more efficient the engine is.

The chamber pressure inside the model rocket engine during the thrust phase will peak at about 225 pounds per square inch.

IGNITION

The temperature required to initiate combustion in a model rocket engine is 550 degrees Fahrenheit. The Solar electrical igniter produced by Estes Industries produces a temperaturer higher than this. The electrical igniter *must be in contact with the propellant* when it functions, or the engine will not ignite. Even a tiny gap between the igniter and the propellant when the igniter functions will prevent the heat produced by the igniter from igniting the engine. Ninety percent of ignition failures are caused by the electrical igniter not touching the propellant when the igniter functions.

EXHAUST

The combustion products produced by a solid-propellant model rocket engine are primarily carbon dioxide and water. The combustion products produced by the smoke-tracking and delay element are primarily carbon dioxide and water, also, but they include a small amount of sulfur dioxide, which produces the characteristic "rotten egg" odor sometimes noticeable around a model rocket after a flight.

The size and shape of the nozzle determine the velocity at which the hot exhaust gases exit the nozzle. The velocity at which these gases exit is important in determining the thrust produced by the rocket engine. The smaller the nozzle, the faster is the velocity at which the rocket's exhaust gases exit the nozzle. The greater the velocity of the exhaust gases, the greater is the thrust produced. The exhaust velocity for most Estes engines is 2,550–2,650 feet per second.

TIME-THRUST CURVES

The shape of the propellant's surface at which combustion takes place is very important in determining the rate at which the hot gases are produced. The burning rate of a solid propellant is the speed at which it undergoes combustion. This is determined partly by the amount of surface area exposed to combustion (see Figure 9–3). The two basic shapes of the solid propellant, usually called the *grain,* are the end-burning grain and the core-burning grain. The core-burning grain has a hole extending into the grain. The core-burning grain has a much larger surface area undergoing combustion than an end-burning grain of the same diameter.

Within about 0.3 second from the instant of engine ignition, this engine has reached peak thrust of 3 pounds (slightly over 13 Newtons). The thrust drops to about 1.3 pounds (6 Newtons) by 0.5 seconds from engine ignition and remains at this level for an additional 0.3 seconds (to 0.8 seconds from ignition). At this time all the propellant has been consumed. After about four more seconds, the ejection charge operates, deploying the recovery device. Since this engine has a "spike" (higher thrust) just after ignition, it is partially core-burning. The constant thrust level is produced by an end-burn of the propellant remaining after the brief core-burn took place.

Nearly all end-burning grains have a small core-burning section so that the initial thrust level is higher than the sustaining thrust level. This can be observed from the time-thrust curves found on pages 98 and 99. This small extra thrust at liftoff is important in safely starting the model rocket on its flight. The slight center coring makes ignition easier, and the high initial thrust quickly accelerates the rocket to a flying speed adequate for the fins to provide aerodynamic stability.

Except for core-burning engines, after the high initial thrust, the rocket's engine changes from a core-burning grain to an end-burning grain with a lower thrust level. The slower burning of the end-burning grain produces a lower thrust level, but the propellant lasts longer (see Figure 9–4).

This engine produces peak thrust (about 5 pounds or about 22 Newtons) at 0.3 seconds after ignition. About 0.6 seconds after ignition, the engine is producing no thrust. All the propellant was consumed in producing the strong "spike" (high thrust) after ignition. The B8-5 engine is a core-burning engine. It causes a rocket to accelerate faster and is useful for lifting heavier payloads, but it has no sustained burn.

A sustained burn at a constant thrust level is very efficient in achieving maximum altitude. (Since the drag is proportional to the square of the velocity, a slower-moving rocket will travel higher for the same expenditure of energy.)

The length of time that the engine burns is partly determined by the

Specific Impulse – 80-83 Lb.-Sec. per Lb.
Exhaust Velocity – 2550-2650 Ft./Sec.

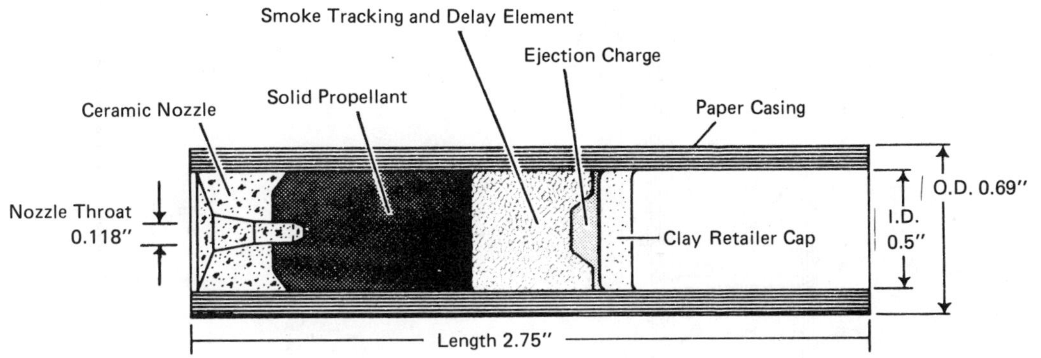

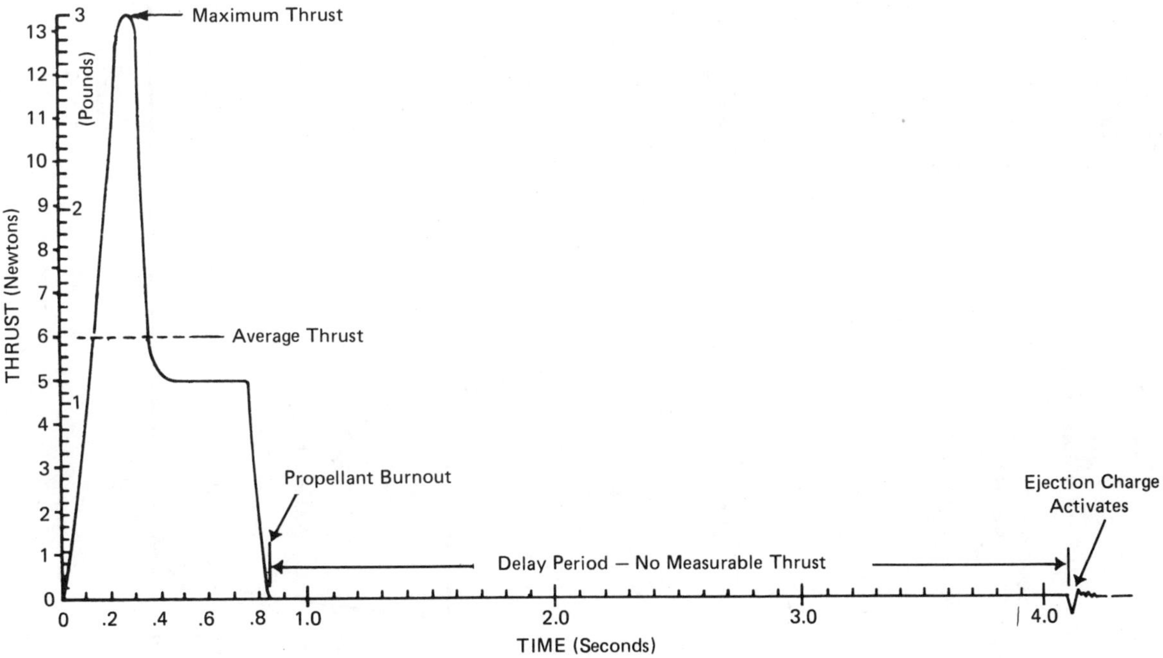

Figure 9-3. Time-thrust curve of Estes B6-4 engine.

Figure 9-4. Time-thrust curve of Estes B8-5 engine.

Exhaust Velocity – 2550-2650 Ft./Sec.
Specific Impulse – 80-83 Lb.-Sec. per Lb.

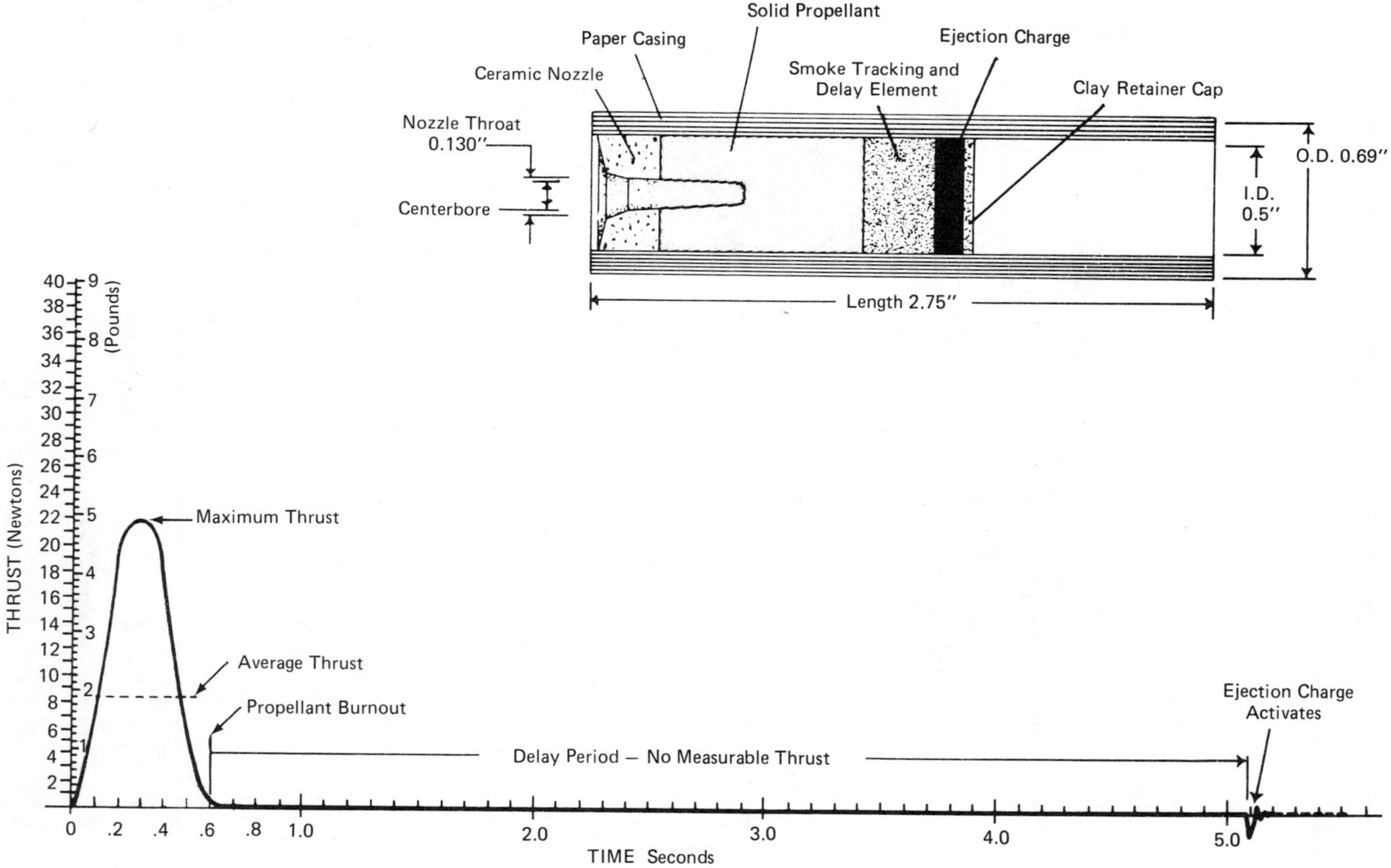

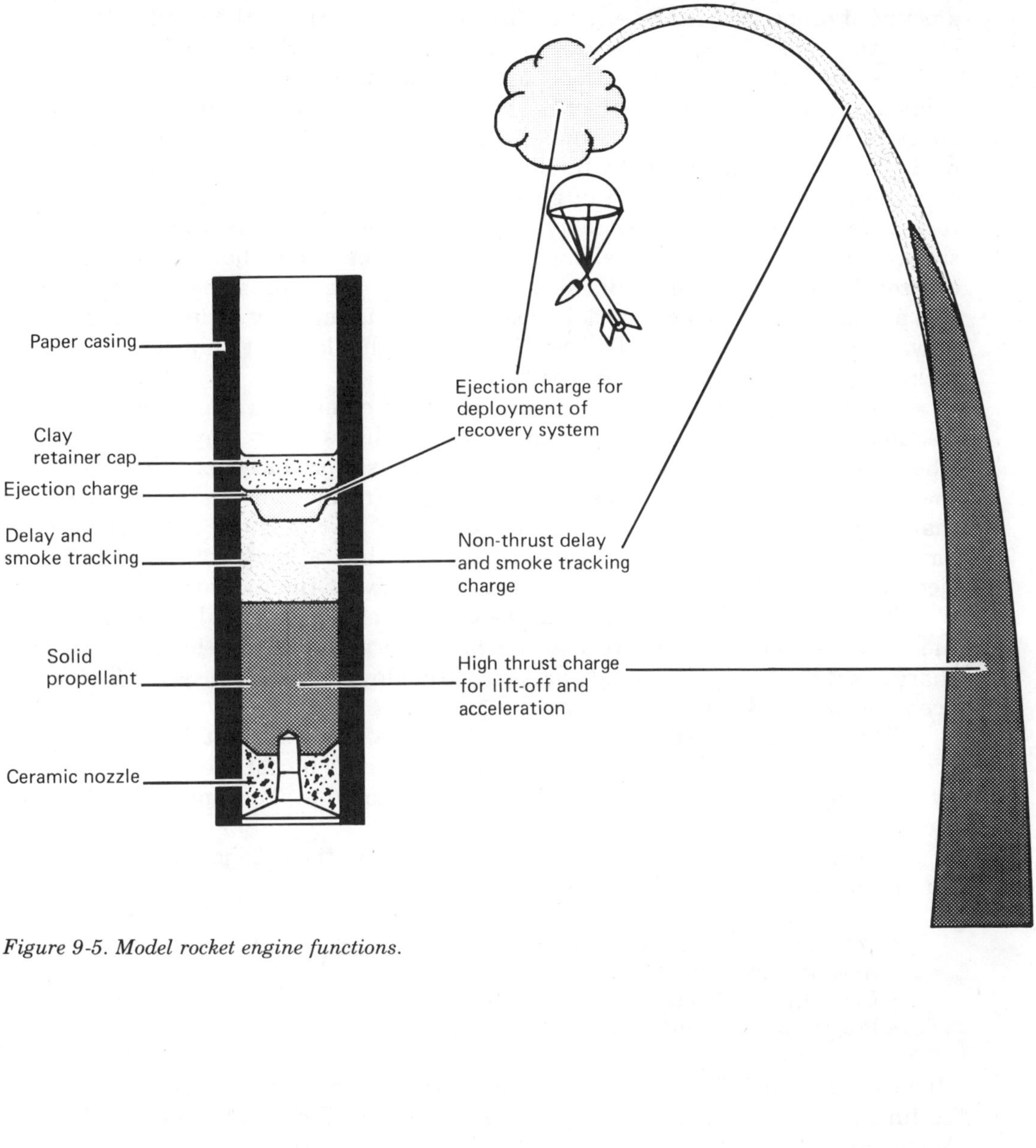

Figure 9-5. Model rocket engine functions.

amount of propellant in the engine. The more propellant in the engine, the longer the "burn" (the length of time the engine's propellant burns) lasts.

As the last of the solid propellant is consumed, the heat ignites the delay and smoke-tracking element. This portion of the engine produces no effective thrust, but it does make smoke. This is important, because the model rocket moving vertically hundreds of feet in the air is very difficult to see. Observers on the ground often see only the smoke trail and not the rocket. The rate at which the delay and smoke-tracking element burns is a steady one. The longer this element is, the longer it will burn. You can control the amount of time this element burns by controlling the length of this portion of the engine. The delay permits the fast-moving rocket to coast upward for a long distance. The ejection charge should function just as the rocket reaches apogee and starts to fall.

As the last of the delay and smoke-tracking element is consumed, the ejection charge activates. The ejection charge produces a very tiny explosion when it undergoes combustion. Part of the hot gases produced exit through the small nozzle. The hot gases are produced by the ejection charge so fast that they cannot all exit through the nozzle, so part of these gases rush forward. These gases moving forward push out the recovery wadding and eject the parachute or streamer, or otherwise activate the recovery system.

The clay retainer cap prevents the ejection charge from falling out of the engine's front end and resists the pressure produced by the ejection charge as it burns until adequate pressure to deploy the recovery system has been developed. See Figure 9–5 on the preceding page.

The average thrust of the engine in Newtons (from the first number in the engine code) can be used to determine the engine's burn time in seconds by dividing the average thrust into the total impulse. The answer is in seconds.

Applying this knowledge to the information on the engine D12-5, you can learn the following about the engine:

- Total impulse: 20 Newton-seconds
- Average thrust: 12 Newtons
- Delay time: 5 seconds
- Propellant burn time: 1.70 seconds

Much additional information about model rocket engines is contained in Technical Note TN-1, *Model Rocket Engines,* and Technical Note TN-2, *Model Rocket Engine Performance.* Both publications are available from Estes Industries.

10
CLASSROOM ACTIVITIES IN SPECIFIC AREAS

MATH AND PHYSICS

★ Model rocketry is a complete miniature technology. The same scientific principles that govern the behavior of full-sized rockets govern the behavior of model rockets.

In earlier chapters we have seen how to use physics to determine the altitudes reached by model rockets, the speeds they reach, the accelerations achieved, and the aerodynamic drag they experience. We have briefly investigated the factors to consider in the design of a model rocket.

In addition, there is much applied mathematics involved in the construction of a model rocket from a kit. Much additional math is involved when a model rocket is designed. Practical math is necessary in the process of selecting and purchasing a kit, a subject that will be considered later in this chapter.

LANGUAGE ARTS

READING FOR MEANING

★ Just being able to say each word correctly does not mean that a youngster is able to derive the proper meaning from a sentence. When involved in the construction of *his* or *her* model rocket, the typical youngster is very involved. The rocket is important. The youngster may just want to get it built so that he or she can fly it, or just to make it the *best* rocket. Either way, the completion of the rocket is important to that youngster.

Each step in the instructions is made up of both one or more illustrations and a text message. The youngster can both *see* and *read* what is to be done. The illustrations are usually so clear that what must be done in that step is clear; however, the written instructions repeat what must be done and add important information. Sometimes the written instructions make that step more clear. An important detail may be presented *only* in the written instructions.

This procedure of presenting information visually in the drawings and verbally in the text makes possible immediate reinforcement of the message. The student can receive the information necessary to do a procedure in two different ways. This can be an important way to reinforce the youngster's confidence in his or her reading ability. If the youngster only looks at the picture, he or she may not build the rocket carefully.

Teachers have frequently commented that specific youngsters finally learned to read on their own to build their own rockets. The teachers refused to help the students, but the students' desires were so strong that they finally took the trouble to read carefully and for meaning.

LAUNCH NARRATION

Any model rocket launch can be more interesting and more exciting with a good narration. Good narrations are essential for demonstration launches.

Naming the rocket and the builder provides basic information for each flight. Giving special information about the rocket or its performance is also possible. Using the jargon of NASA launches can add much to the "atmosphere" of a launch.

When launching scale models, a whole new arena of knowledge is opened up. Relating facts about the rocket that has been scaled can add much interesting information to the narration. Sometimes items of local interest can be added, such as if a local company helped manufacture parts for that particular rocket. For launches held in honor of a specific space shuttle flight, you may find that one of the astronauts came from your town or once lived in your area, or that one of the payloads was designed or built in your state. Include the launch date and time, the duration of the mission, the payload, and the anticipated touchdown date, time, and place.

The type of narration you use will be determined by the type of launch. For a class or school launch, perhaps the fastest narration possible will be best in the interest of getting all the rockets successfully launched in a minimum amount of time. For a Goddard Day launch, present facts about Robert Hutchings Goddard and his important discoveries.

For any narration, keep it moving! The pace must be brisk. Dragging out the launch with slow-paced, boring, or unnecessary narration will cause your audience to lose patience, and you to lose your audience.

REPORTS ON LAUNCHES

Reports on launches can be made for a variety of reasons. The report can be an activity to develop skills in accurate reporting of facts. This report can be a written report, or it can be a verbal activity to help develop skills in public speaking as well as in gathering and presenting facts concisely.

Reports on model rocket launches make excellent reading in school newspapers. Not only do students experience the learning involved in writing their reports, but they also have the opportunity to get their reports published. As a real challenge for a good student or a good group, let them develop the entire launch as a radio report.

Students (and teachers!) seldom get enough ego-boosting and gratification for a job well done. Don't hesitate to "wave your own flag." Anyway, principals, parents, school boards, and the community like to hear about things the school is doing well.

CREATIVE WRITING

The opportunities for using model rocket activities as worthwhile learning experiences in language arts are limited only by your imagination. The creative activities that can focus on model rocketry include writing poems. Even the haiku form of poetry can be written about high-technology model rocketry. Writing stories about model rocket activities is also an excellent creative-writing activity.

CONSUMER EDUCATION

Building and launching a model rocket provides an excellent opportunity to learn how to be a better consumer. We all spend much of our time and virtually all our money as consumers. An experience in model rocketry gives your students a chance to develop and exercise several consumer skills that will always be useful.

Once a student has decided to purchase and fly a model rocket, several other decisions must be made. The type of rocket to buy and the type of kit to buy are two decisions for the youngsters. Learning to read and understand the model rocketry catalog becomes very important. This may be the first time that the student has had the opportunity to make a purchasing decision from a catalog. A number of factors enter into the decision of which kit to buy.

The order blank must be properly filled out and mailed with the proper payment. This activity is not as easy as it looks, especially to a first-time mail-order customer.

PREPARING THE ORDER

The easiest way for you, as a teacher or club leader, to place an order is to require everyone to purchase the same rocket and engine. This way you only have to determine the cost to each student and collect it. Not only are the mechanics and the mathematics of placing an order made easy, but you will not have to worry about trying to help students who are building different types of model rockets. The maximum learning value of a model rocketry experience takes place when the youngsters are presented with the maximum number of valid learning activities that they successfully complete.

The factors that may enter into the decision of which model rocket to buy, build, and fly include at least the following:

- determination of how much the individual can spend (a basic decision involving "desire" versus "reality");
- determination of the amount of discount, if any, the order will receive;
- deciding whether the order will be placed by mail or carried to a local store (analyze the advantages and disadvantages of each method);
- selecting a kit at the appropriate skill level;

- choosing the appropriate engine for the rocket (involves selecting an engine from those recommended considering the size of the launch site, considering the usual winds, and awareness of the amount of money available).

The student may need help in considering one or more of these factors. Additional considerations that you should help with include determination of cost of postage, amount to be paid as a handling charge, amount to allow for taxes, how to safely transport the money to school, and how to make change when necessary. These may seem to be trivial considerations, but they aren't, especially for younger students.

Unless your class includes some experienced model rocketeers, you should insist that all rockets purchased be skill level 1 models. Even if you teach a class for the gifted, these youngsters, while academically talented, may lack the basic coordination necessary for successful construction of higher skill level kits.

WRITING THE ORDER

If each youngster is to place his or her own order, make enough copies of the order form for each youngster to have one, plus a few extras. If the order is to be placed as a class order, make enough copies of either the individual order form or the school and club order form for each youngster to have one, plus a few extras.

For maximum learning value, have each student prepare the order form completely. The name and address portion of the order form should be filled out. If you are using an individual order form, now is a good time to complete the miscellaneous questions asked on the order form and on the order envelope. Leave blank any spaces set aside "for office use only."

Next, the name of the rocket kit, identification of the engines, and other items ordered are entered. Also enter the "cat. no." and quantity information correctly in the proper columns. Next enter the proper current price of each item in the "unit price" or "price ea." column. Then extend each horizontal line by entering the proper "total" in the proper column.

Remind the students that model rocket engines usually come in packages of three for standard and D engines and packages of four for mini-engines.

After the basic order is completed to this point, add up the "total" column to determine the amount owed for the merchandise ordered. Complete the rest of the information required at the bottom of the order form to determine the total amount owed, including taxes, shipping and handling, and so on. Then recheck all items on every order.

If each student is to mail in an individual order or to take the completed order form to a local store, each student can now take care of getting his or her order filled. If the orders are mailed, make certain that each student puts in a check or money order (*not cash*) of the correct total amount in the envelope with the completed order form and mails it. If the orders are to be filled through a local store, please alert the store well in advance. Most store owners are quite happy to cooperate. If the orders are to be combined as a class order, as is usually the case, the individual orders must be combined into one group order.

Be certain to study the terms of sale on the order form very carefully. Sometimes you can save money on postage by prepaying the order. Sometimes the discount will vary, depending on the total amount of the order and/or whether or not the order is prepaid.

Always allow plenty of time for the order to reach the company to which it is mailed, for the order to be filled, and for the merchandise to reach you. Allow extra time for weekends, holidays, peak mail periods (such as at Christmas), and during heavy demand seasons (such as late in the school year). Place the order early to avoid disappointment.

Check the order as soon as it is received to make certain that all items ordered were delivered and to be certain that the packing list with the order is accurate. Note any back-ordered items.

At this point your students have had a meaningful experience in consumer economics. They have had practice in reading a catalog for information, reaching practical purchase decisions based on practicality as well as on desire, properly completing an order form, making change, anticipating needs and allowing for delays, and successfully completing a mail-order transaction. And the best, building and flying their own model rockets, is yet to come!

GENERAL SCIENCE

★ The real science inherent in model rocketry makes it ideal for science classes.

The concepts studied may be very simple, or they may be quite sophisticated. Select concepts that you want your students to learn. Choose activities appropriate to helping your students understand these concepts.

The variety of meaningful learning experiences in model rocketry is virtually limitless. Model rocketry provides effective learning experiences for classes from the fourth grade (and sometimes younger students, under the right conditions) up to college classes. The high motivation involved in model rocketry makes it useful for special-education classes. The open-endedness of the varied learning experiences possible makes it ideal for classes for the gifted.

Use the appropriate sections of this book and the suitable publications available from the various manufacturers and other references in selecting specific learning activities.

PREDICTING PERFORMANCE

Accurate prediction of a model rocket's performance requires a detailed knowledge of the rocket, the flying conditions, and the engines to be used.

If you are lucky enough to have a proven computer program that will provide accurate predictions of the height that a given model rocket will achieve with a specified engine, just input the proper data. There is another way, however, to determine a close approximation of the actual performance before the flight. This involves careful use of Technical Report TR-10, *Altitude Prediction Charts,* published by Estes Industries.

This publication was developed to help rocketeers who wanted to predict the performance of their model rockets but who did not have access to good computers and the necessary software to perform the calculations for them. Determining the probable altitude that the rocket will reach using TR-10 requires knowledge of the exact diameter of the rocket, the weight of the rocket without engine, the engine to be used, a good approximation of the C_D (aerodynamic drag coefficient) of the model rocket, the catalog of the manufacturer who made the engine, and Technical Report TR-10.

While the process may initially appear quite difficult, the complex formulas have already been applied through the use of sophisticated computer programs to produce graphs that are read to save the user from having to use the complex formulas. The procedure requires only eight simple steps to determine the predicted altitude of a given flight and provides the probable coasting time for that flight, so that an engine with the proper time delay can be chosen to achieve parachute ejection at apogee. TR-10 allows you to adjust the predicted altitude to allow for the actual air temperature at the time of launch and the elevation above sea level.

The accuracy of prediction achieved by this method can be further refined by tracking the flight and using the actual result achieved. This allows the aerodynamic drag coefficient you assumed for the rocket to be altered in predicting the altitude to be reached on subsequent flights.

If the eight-step procedure seems too lengthy, a simple slide-rule type calculator called the Estes Predicted Altitude Computer permits you to predict the height to be reached on a given flight through a few simple calculations and movements of the slide. However, students will gain little insight into the factors involved in altitude performance through the use of this calculating device.

DESIGNING MODEL ROCKETS FOR SPECIFIC MISSIONS

Decide exactly what you want to achieve on a specific flight. Base your decisions of which model rocket and model rocket engine to use on the mission objective.

If you want to achieve maximum height, select a model rocket with very low drag. Usually, you will select a model rocket with very light weight. This may not always be true, however. As you will learn when using TR-10 to predict the height to be reached, sometimes a slightly heavier model rocket will have a thrusting ballistic coefficient that allows the rocket to coast higher during the coast phase of the flight and thus reach a slightly higher altitude than will be reached by a similar-sized rocket with a lighter weight.

If the mission is to achieve maximum flight duration, then the rocket should be as streamlined as possible. Usually, it should also be as light as possible. The recovery system, whether parachute or streamer, should be as large as possible to maximize the duration of the flight's recovery phase.

For payload launches, the payload compartment must be large enough to properly contain the payload.

Once you have a rocket that successfully accomplishes the mission, you may be able to improve its performance by slight modifications.

INDUSTRIAL ARTS

Industrial arts teachers have widely adopted model rocketry as part of their curricula. This is especially true at the elementary and junior high-school levels, where most industrial arts programs include model rocketry.

While the objectives for using model rocketry in the curricula for industrial arts teachers are often different from the objectives established by science teachers, model rocketry makes a meaningful contribution to the educational process. The high-technology aspects of model rocketry are emphasized in some curricula (especially those on power technology) and minimized in other curricula. Just as for science teachers, each industrial arts teacher uses model rocketry to help his or her students accomplish their own objectives.

Many industrial arts teachers and science teachers cooperate in presenting a unit on model rocketry. While the objectives of each teacher are different, the material can be successfully presented to both groups simultaneously. In addition to benefiting from the exposure to the different knowledge and attitudes emphasized by each teacher, the students can benefit by sharing efforts. The industrial arts students can gain a greater appreciation and understanding for the scientific principles governing the operation of the model rockets, and the science students can gain a greater respect for the practical, hands-on approach of the industrial arts class. Too often, we teachers segregate knowledge into artificial compartments to the detriment of the intellectual growth of the students.

Many industrial arts classes actually design and then build their own launch pads and launch controllers. These are valid learning experiences. Not only do these projects result in functioning devices, sometimes at lower cost than use of the standard kits, but the students usually achieve a greater understanding of both the theory and the mechanics of these devices.

Construction of a multiple launch system is a challenging project. However, it is achievable by an upper elementary school or older class. These devices are often objects of pride for those involved in their design and construction. They are great for class launches and demonstrations, too.

If one of these multiple pad launchers is built, a few suggestions are in order. Be sure to have a very good power supply. Use of an automobile battery (either in the car or removed from it) or a motorcycle battery with a good charge is recommended. Use heavy (low-resistance wire) for all wiring. Construct two launch pad sets, if possible. Use one for the actual launches while the other is being readied with the next set of model rockets to be launched.

The Range Safety Officer or the appointed deputy should check all rockets before any are placed on the launch rods, as well as after the micro-clips are attached. Always disconnect the launch pad set on which the rockets are being readied as a safety measure while the other launch pad set is being used, even if each launch pad is wired separately. Always use the standard wiring pattern, which provides for a continuity light, a safety key, and a launch button with a spring return. Make certain that the two pad sets stay at least 20 feet from each other and from the launch control panel.

INDIVIDUAL RESPONSIBILITY

Deciding to buy a model rocket kit and selecting the exact kit involve individual responsibility.

Building the kit is totally an individual responsibility. The appearance of the kit and its performance are determined by the care exercised in the construction of the kit. The student cannot blame anyone else for poor construction or poor performance of his or her rocket. This places the entire responsibility for the rocket on the individual. This often makes a student stop and think and perhaps try harder to do his or her best.

In addition to the individual responsibility inherent in the process of purchasing the model rocket kit and building it, there are many opportunities for student growth in responsibility in model rocket activities.

The model rocket must be prepared for launch in a certain way. This prepping is entirely up to the student. Any failures to follow directions accurately can result in embarrassing failures in performance! This brings peer pressure into play very effectively.

If the launch is conducted in a formal manner with the Launch Control

Officer, Range Safety Officer, Recovery Crew, Tracking Team, and Data Reduction Team, the opportunity to learn and successfully perform a series of tasks necessary to a successful launch is immediately presented. Each student who performs one of these functions is immediately responsible for performing a series of necessary tasks. He or she is necessary to the others. Each of these functions requires specific skills and specific tasks. Each student must perform his or her function well for the launch to proceed smoothly. Everyone depends on everyone else. Any foul-ups will become immediately apparent to the others involved. This is a lesson that some youngsters are slow to learn. Again peer pressure works to reinforce the importance of each person being responsible for his or her actions.

Rotating the youngsters between the assigned functions is a good way for them to learn to appreciate the responsibilities of each position as well as learning the tasks necessary for each function. In fact, we have found that youngsters, once properly trained, are usually more strict with each other in the enforcement of safety and other rules than an adult leader. This is not just because of the opportunity presented to exercise their "power," but also because the model rocket activities are very important to them.

AEROSPACE EDUCATION

This subject is one that holds great attraction for students of all ages and all levels of academic achievement. It involves the study of all aspects of flight and space travel.

A large number of specific learning activities are possible in aerospace education. Model rocketry is only one.

Aerospace education has much to offer to every teacher. Specific activities can be used in almost any unit on any subject. We recommend that you attend one or more of the aerospace education workshops presented at universities around the country each summer. You will find the experience personally fascinating and professionally rewarding. Also, many of these universities will come to your school district and offer the course for your district and/or offer the class as an evening or Saturday class. Contact your local university or your director of aerospace education for your regional Civil Air Patrol office.

The National Aeronautics and Space Administration (NASA) regional offices can offer help in answering specific questions and in supplying excellent literature. While currently operating on severely limited funds for its aerospace education program, NASA may be able to schedule one of its "Spacemobiles" to visit your school. The programs presented by the speakers who operate these mobile programs are excellent for school assemblies, individual classes, presentations to parents, and local television programs. Try to schedule one of these presentations now.

Kites offer excellent opportunities for learning. The "Learning with Kites Program" by Estes Industries and Hi-Flier provides an excellent series of self-teaching learning experiences.

Student-made "helicopters," flying paper airplanes, and similar devices offter great learning potential. Use your imagination and don't be afraid to experiment! Your students will love you for it.

MOTIVATION

The motivation of the students—no matter what level—is usually unbelievably high when dealing with model rockets. *Use* this motivation to help them learn to read, to follow directions exactly (even when this isn't the easiest way), to use some basic mathematics, to cooperate effectively, to accept individual responsibility, and to learn and use as much science as you have time for.

11 ADDITIONAL ACTIVITIES

Model rocketry lends itself to many types of group activities and individual projects. Some of these have been covered in preceding chapters. Any project involving model rocketry can have educational benefits and encourage personal growth as well.

CONTESTS

★ Contests provide opportunities for rocketeers to exhibit their expertise.

Model rocket contests are perhaps the most interesting type of group activity involving this hobby. There is more than a little attraction in competition for most students. Model rocketry does not require special skills or talents, and does not demand long hours of training (as do conventional competitive sports), so everyone starts out as an "equal" in model rocketry competition.

This is not to say that model rocket competition does not require knowledge and skills. The knowledge and skills required, however, can be acquired and developed by almost anyone, with a little effort.

Model rocket contests can serve as a learning incentive because they encourage students to learn as much about the theory and design of rockets as possible in order to gain a competitive "edge." In the end, competition becomes as much a test of the students' knowledge as a test of their building and flying skills. The more complete a student's knowledge, the better he or she is able to accomplish challenging projects successfully. (See Figure 11–1.)

The basic areas of competition for model rocketry are altitude, payload handling, duration, spot landing, and craftsmanship. A number of duration events include gliders, which are launched as rocket-boosted vehicles, part or all of which return by aerodynamic means.

ALTITUDE EVENTS

★ In altitude competition, the contestants attempt to send their model rockets to the highest possible altitude.

Rockets flown against one another in this type of competition must each use the same type of standard model rocket engine, and may be required to carry a standard 1-ounce payload. The payload is usually a miniature "sandbag" carried in the payload compartment or in a hollowed-out nose cone.

The rockets in altitude competition can be flown in various "classes," categorized by the type of engine used to power the rocket. There may also

Figure 11-1. Spectators.

Figure 11-2. Commanche-3 on launch pad.

be separate competition classes for multiple-stage rockets. (See Figure 11–2.)

Altitude competitions imply a means for determining the maximum altitude attained by the rockets. The altitudes can, of course, be determined by using a theodolite, a commercial device such as the Estes AltiTrak, or a homemade device. The rockets should be flown one at a time, and they should be required to make a qualifying flight (a flight with a successful and safe launch and recovery, and that has been successfully tracked).

DRAG RACE

★ Drag races are unusual and exciting contests.

A model rocket "Drag Race" is one of the more exciting competition events. In this event, two model rockets are launched simultaneously. A point system is used to determine the winner of the event. One point is awarded for the rocket that achieves *first motion* at launch. One point is awarded for the rocket reaching the *lowest altitude.* One point is awarded for the rocket *reaching the ground last.* As with other events, the rockets must have a safe launch, flight, and recovery.

EGG LOFT

★ The purpose of the competition is to successfully and safely carry a fresh hen's egg to the highest altitude and recover it unbroken.

The "Egg Loft" event requires skill in design as well as in building, flight planning, and overall preparation. The usual entry in the egg loft event is a specially designed rocket or a modified kit. A specially designed payload section, with padding, is required.

Since egg loft rockets tend to be rather heavy, an engine or engines with a high initial thrust is the best choice. Heavier rockets require a "kick" at liftoff to get them moving quickly to a speed at which the fins stabilize them. Of course, the high initial thrust means that the thrust duration of the engine will be brief, and this means that there is less sustaining thrust to boost the rocket to altitude. All B engines have the same *total* thrust, 5 Newton-seconds. The thrust duration of a B4 engine is 1.2 seconds, while that of a B8 engine is 0.6 seconds. Because of this, the modeler should be especially mindful of the drag created by the rocket, because rockets with a lot of drag will not coast as far as those with little drag. This also has an effect on the selection of the time delay between propellant burnout and activation of the ejection charge.

The recovery system is critical to success in egg loft events because recovery system failure almost always means a heavy landing and a cracked or crushed payload.

Altogether, egg lofting is the most challenging of competition events because nearly every aspect of building, preparation, and flying is critical to success.

ALTITUDE PREDICTION

★ Altitude prediction is an event in which the goal is to most closely predict the altitude reached by a rocket on a specific flight.

The modeler does not necessarily have to achieve the highest altitude, but to predict the apogee (maximum altitude) of his or her rocket as closely as possible. Using knowledge of aerodynamics, physics, engine performance, and flight characteristics, which can be gained through study of topics covered in Chapters 4 through 6 of this book, a modeler can quite often predict the apogee of his or her rocket within a few feet (see Figure 11–3).

Figure 11-3. Calculating.

As with any altitude event, altitude prediction requires a means of determining the maximum altitude of the rocket. The degree of difficulty of the event may be increased by requiring that qualifying predictions must be within a certain percentage of error.

DURATION

★ Duration events can be flown in any engine class, and may be broken down into three types: parachute duration, streamer duration, and glider duration.

Duration events must be timed from the instant of first motion on the launch pad to the moment of touchdown on recovery. (Models are not to be caught, but must be allowed to touch the ground on their own.) In addition to the requirements of a safe flight and recovery, rockets flown in a duration event must touch down in sight of the judges and timers.

Model rockets designed for duration competition are generally built as light as possible, with recovery systems with large surface areas. Low drag is an asset to a duration model because, in general, a higher apogee means a longer period of time in the air.

SPOT LANDING

★ Spot landing competition requires that the contestants' rockets touch down as close as possible to a marker placed in an arbitrary location on the ground.

Spot landing models usually use featherweight, tumble, or streamer recovery so that there will be little or no lateral drift during descent. To add a little fun to the event, the spot landing marker may be a rocket that has been launched first and allowed to land where it may. The measurements are usually made from the tip of the rocket's nose cone to the marker.

CRAFTSMANSHIP

★ Craftsmanship events require the rocketeer to use design and building skills to create a model rocket that is as "clean" in appearance as possible. Fabrication, gluing, finishing, and painting skills are required for success in this type of event.

SCALE MODELING

★ The goal in scale modeling is to build a model that is a minature of an existing "real" rocket or spacecraft, either current or historic.

Scale modeling includes elements of craftsmanship events and also tests the rocketeer's research and engineering skills. The scale model should be as near perfect in its proportions, markings, and performance as possible. Documentation may be required, and it may include photographs of the prototype rocket to substantiate shape, size, coloring, and markings, as well as text descriptions and flight descriptions.

Scale modeling events can be combined with other events, such as altitude or duration, to create special categories of competition. Even if the scale event is not combined with other types of competition, the entries in scale competition should be required to complete a successful and safe launch, flight, and recovery.

It should be stressed that any type of competition will require that the models complete a successful flight, which should include as vertical a flight as possible and a safe recovery of all parts with minimal damage.

DEMONSTRATIONS

★ A carefully planned and timed "demo" (demonstration) will yield benefits to a club in increased membership and public awareness. For a school, hobby shop, or manufacturer sponsor, the demo will be excellent public relations. Demonstrations help the hobby to become known as the safe and fun activity that it is (see Figure 11–4).

You will, of course, need a sponsor for the demo. If the club is putting on a demo for a school, the school will be able to provide the space for flying. A hobby-shop owner can help the club obtain permission to use a shopping center parking lot or a park near the hobby-shop location for the demo. It is

Figure 11-4. Demonstration.

important that the demo be held near the hobby shop so that those spectators interested can obtain information and/or rocketry materials. Part of the reason a hobby shop will sponsor a demo is to increase its rocketry business. In some cases, a hobby shop will provide advance publicity, and perhaps engines and other supplies for the demo.

If you want manufacturer support for a demo, it's best to write to the manufacturer with information about where and when you want to hold the demo. Also, let the manufacturer know the name and address of the hobby shop. The manufacturer may arrange a demo with a hobby-shop owner for you and may provide some supplies and advance publicity suggestions.

You should arrange for local newspapers to have reporters on hand for the demo, as well as radio and TV stations. Contact the publicity departments of local radio and TV stations, by phone or letter, and let them know that you will be conducting a model rocketry demonstration. Be sure to provide details as to where and when the demo will be held, along with some information on just what model rocketry is. Be sure to give the name, address, and phone number of your club advisor, senior member in charge of publicity, or responsible officer for the newspaper or TV or radio station to contact for more information.

You will want to conduct public relations yourself during the demo. Give a brief but interesting and informative narrative for each launch. As stated earlier, you should have several people available to answer questions, and you should also make available model rocketry pamphlets, flyers, and catalogs, which you can obtain through the hobby shop or a manufacturer.

ADDITIONAL STUDENT ACTIVITIES

In a classroom situation model rocketry is great for promoting cooperation. If you have assigned the various jobs required for safe operation of a launch range (Range Safety Officer, Launch Officer, Recovery Crew, and so on), your students will learn quite a bit about working together to achieve success in a group effort.

The study of model rocketry overlaps with the study of space exploration in many areas. Model rocketry can be used as an introduction to the history of space exploration, space technology of the future, and even alien life forms. Some activities that may be undertaken as enrichment activities are designing spacecraft for voyages to other stars, designing a hypothetical alien spacecraft, or even designing and building a model of a space habitat.

School assemblies, especially when given during "field days" near the end of the school year, offer a great opportunity for students to exhibit their knowledge of rocketry and allied topics, through verbal and written presentations as well as flying demonstrations.

The classroom applications of model rocketry are limited only by the imaginations of the students and teachers.

SCIENCE FAIR PROJECTS

Numerous possibilities for projects are available in model rocketry. The project may be simple, such as a presentation in posters of a model rocket's flight profile. More elaborate projects can also be considered, such as testing the drag characteristics of different nose cones or optimizing performance of a specific design through experimentation by alteration of weight achieved through hollowing out the nose cone or sanding fins to reduce weight and adding weight to the nose cone to increase weight.

Science Fair projects should be good learning experiences for students to try to do better than their previous best. The projects should make the students feel good. Competition is acceptable and is a basic part of most science fairs, but no student should feel that he or she "failed." Development of the project and participation in the activity should be an accomplishment the student feels proud about.

The "scientific method" should be used in the development of all experiments and the reports on these experiments. Careful record keeping is essential. The use of controls is appropriate for projects that involve research. Whether the research undertaken has been done before is not the important point. The research must be carefully planned, faithfully executed, precisely recorded, and the resulting data thoughtfully analyzed and reported. Any results determined must be fully documented with valid evidence.

Much more information regarding science fair projects and suggestions for about 100 projects is contained in the Estes Industries publication *Projects in Model Rocketry*.

CLUBS

Model rocketry is a lot of fun and a great learning activity when undertaken alone. It is even more fun when done with others. Model rocket activities at school often result in the formation of model rocket clubs.

It is important that model rocketry be kept in proper perspective in a school environment. It is a great teaching and learning aid. It is fun for all those involved, and it offers a great spectator sport for almost everyone else. While model rocketry has proved to be extremely effective for motivating and promoting success in special education classes and has been used heavily in other classes (even as much as half of the course for precollege high-school physics), it is only one aspect of the total school program. The appeal of model rocketry is so strong that many science clubs that undertake model rocketry activities end up as model rocket clubs. This is great for those who are "gung-ho" rocketeers, but is not fair to the others in the club. Sometimes the only solution is to have two clubs, a regular science club and a model rocket club. Many excellent suggestions for organizing and running a model rocket club are contained in the Estes publication *Guide for Aerospace Clubs*.

COOPERATION SKILLS

Because of its characteristics as a miniature technology, model rocketry offers unsurpassed opportunities for working together. While model rocketry is often engaged in as an individual activity, it often involves many people. A large launch requires the successful cooperation of many people performing specific tasks in competent and timely fashion. This necessary interaction provides excellent learning opportunities. Not only do the students see the necessity of cooperation, they *experience* it!

APPENDIXES

APPENDIX A: BIBLIOGRAPHY AND RECOMENDED READING

ROCKETRY AND SPACE TRAVEL

Baker, David. *The Rocket: The History and Development of Rocket and Missile Technology*. New York: Crown, 1978.

______. *Space Shuttle*. New York: Crown, 1979.

Cowley, Stewart. *Spacecraft 2000 to 2100 AD*. New York: Crown, 1979.

Freeman, Michael. *Space Traveller's Handbook*. New York: Sovereign Books, 1979.

Glasstone, Samuel. *Sourcebook on the Space Sciences*. Princeton, N.J.: D. Van Nostrand, 1965.

Hirsch, David, and Howard Zimmerman. *Spaceships*. New York: Starlog Press, 1980.

Malone, Robert. *Rocketship*. New York: Push Pin Press, 1977.

Ronan, Colin (editor). *Encyclopedia of Astronomy*. Secaucus, NJ: Chartwell Books, 1978.

MODEL ROCKETRY

Banks, Michael. *Second Stage: Advanced Model Rocketry*. Milwaukee, WI: Kalmbach Books, 1985.

Boles, Dane. *Guide for Aerospace Clubs*. Penrose, CO: Estes Industries, 1971.

Cannon, Robert. *Elementary Mathematics and Model Rocketry* (second edition), Technical Note TN-5. Penrose, CO: Estes Industries, 1979.

______. *The Fine Art of Payload Launching*. Technical Note TN-4. Penrose, CO: Estes Industries, 1978.

______. *Is That Parachute Too Big?*, Technical Note TN-3. Penrose, CO: Estes Industries, 1977.

______. *The Laws of Motion and Model Rocketry*. Penrose, CO: Estes Industries, 1979.

______. *Model Rocket Contest Guide*. Penrose, CO: Estes Industries, 1968.

______. *Model Rocket Launch Systems* (Second edition). Penrose, CO: Estes Industries, 1976.

______. *Projects in Model Rocketry*. Penrose, CO: Estes Industries, 1974.

Eisenberg, Robert and Wendall Hyde. *Countdown: Skydiver, Rocket, and Satellite Motion on Programmable Calculators*. Forest Grove, OR: Dilithium Press, 1979.

Estes, Vernon. *Building a Wind Tunnel,* Technical Report TR-5. Penrose, CO: Estes Industries, 1965.

______. *Multi-Staging,* Technical Report TR-2. Penrose, CO: Estes Industries, 1963.

______. *Rocket Stability,* Technical Report TR-1. Penrose, CO: Estes Industries, 1963.

Gregorek, Gerald. *Aerodynamic Drag of Model Rockets,* Technical Report TR-11. Penrose, CO: Estes Industries, 1970.

Hajek, Stanley, and Raymond Schuette. *Space Age Technology*. Penrose, CO: Estes Industries, 1970.

______. *Teacher's Manual for Space Age Technology*. Penrose, CO: Estes Industries, 1970.

Kellner, Wayne. *Recovery Techniques,* Technical Note TN-6. Penrose, CO: Estes Industries, 1973, 1978.

Mandell, Gordon. *Front Engine Boost Gliders,* Technical Report TR-7. Penrose, CO: Estes Industries, 1964.

______. *Rear Engine Boost Gliders,* Technical Report TR-4. Penrose, CO: Estes Industries, 1963.

Mandell, Gordon; Caporaso, George; and Bengen, William. *Topics in Advanced Model Rocketry*. Smithtown, NY: (W. Bengen, 10 Ursular Court, Smithtown, NY 11781), 1973.

National Association of Rocketry. *United States Model Rocketry Sporting Code*. Elizabeth, PA: 182 Madison Drive, 1979 (current edition).

O'Hara, John. *RECROC—Recreational Rocketry*. Penrose, CO: Estes Industries, 1978.

Olney, Ross. *Out to Launch: Model Rockets from Design to Launch*. New York: Lathrop, Lee, and Shepherd, 1979.

Pratt, Douglas. *Basics of Model Rocketry*. Milwaukee, WI: Kalmbach Books, 1980.

Saltrick, Daniel, Alfred Kubota, and Robert Cannon. *Aerospace Education and Model Rocketry* (second edition). Penrose, CO: Estes Industries, 1975.

Schall, William. *COUNTDOWN: Mathematics and Model Rocketry*. Penrose, CO: Estes Industries, 1981.

Smith, Harry, and Henry Warden. *Industrial Arts Teacher's Manual for Model Rocketry*. Penrose, CO: Estes Industries, 1980.

Stine, G. Harry. *Handbook of Model Rocketry* (fifth edition). New York: Arco Publishing, 1983.

Alpha Book of Model Rocketry (second edition). Penrose, CO: Estes Industries, 1976.

Altitude Tracking, Technical Report TR-3. Penrose, CO: Estes Industries, 1963.

Cluster Techniques, Technical Report TR-6. Penrose, CO: Estes Industries, 1967.

Model Rocket Engines, Technical Note TN-1. Penrose, CO: Estes Industries, 1972.

Model Rocket Altitude Prediction Charts (second edition) Technical Report TR-10. Penrose, CO: Estes Industries, 1971.

Model Rocketry Study Guide. Technical Report TR-8. Penrose, CO: Estes Industries, 1972.

MODEL BUILDING

Edmondson, Harold. *Famous Spaceships of Fact and Fantasy . . . and How to Model Them*. Milwaukee, WI: Kalmbach Books, 1979.

Schleicher, Robert. *The ETV Model Book*. Radnor, PA: Chilton Book Company, 1979.

MOTION PICTURES

Model Rocketry: The Last Frontier. Ramsgate Productions, available from Southerby Productions, 5000 East Anaheim St., Long Beach, CA 90804.

OTHER USEFUL PUBLICATIONS

Service, Robert. *Space Exploration* (1976 revision). Boy Scouts of America Merit Badge Guide #3354, Boy Scouts of America, Dallas/Ft. Worth Airport, TX, 1976.

______. *Model Aircraft Laboratory*. CAPP33, Civil Air Patrol, Maxwell Air Force Base, AL.

APPENDIX B: SOURCES OF KITS, SUPPLIES, AND EQUIPMENT

MANUFACTURERS

A & A Engineering
P. O. Box 6284
Buena Park, CA 90620
(Manufacturer of electronic launch controllers)

Aerotech
10869 Paiute Way
Rancho Cordova, CA 95670
(Manufacturer of high-power model rocket engines)

Competition Chutes
c/o Joan Kuczek
49354 Bay Lane
New Baltimore, MI 48047
(Manufacturer of fabric parachutes, stock and custom. Catalog: $1.00)

Competition Model Rockets (CMR)
Box 7022
Alexandria, VA 22307
(Manufacturer of kits, supplies, parts, and custom competition devices. Catalog: $.50)

Estes Industries
Dept. 217
Penrose, CO 81240
(Manufacturer of a full line of model rocket kits, engines, parts, supplies, and publications. Catalog: $1.00)

Flight Systems, Inc. (FSI)
9300 East 68th St.
Raytown, MO 64133
(Specialized manufacturer of high-power kits and engines. Catalog: $1.50)

National Association of Rocketry Technical Services (NARTS)
182 Madison Drive
Elizabeth, PA 15037
(Publishes technical information and scale modeling information packets. NAR members receive discount.)

Pine Cap Associates
Box 2470
Fort Walton Beach, FL 32549
(Manufacturer of kits. Catalog: $1.00 in stamps)

DISTRIBUTORS

AAA Model Aviation Fuels
RD 6, Box 172
Edwards Avenue
Clarks Summit, PA 18411
(Distributor of high-power kits, parts, engines, and accessories.)

Belleville Wholesale Hobbies
1827 North Charles St.
Belleville, IL 62221
(Distributor of model rocket kits, parts, engines, supplies, and accessories.)

Celestial Enterprises
4412 Crabapple St.
Ft. Worth, TX 76137
(Distributor of kits and parts.)

Discount Hobbies
113 Midland Ave.
Kearny, NJ 07032
(Distributor of model rocket kits, parts, engines, supplies, accessories, and plastic model kits.)

Pinecap Associates
Box 2470
Fort Walton Beach, FL 32549
(Distributor of high-power kits and engines.)

Rocket Enterprises
17 Kirkbridge Rd.
Flemington, NJ 08822
(Distributor of model rocket kits, engines, end equipment.)

APPENDIX C: ORGANIZATIONS

Aerospace Education Association
1910 Association Drive
Reston, VA 22091

British Space Modeling Association
15 High Street
Ditchling
Sussex BN6 8SY
England

Federation Aeronautique Internationale
Paris, France

National Association of Rocketry
182 Madison Drive
Elizabeth, PA 15037

L-5 Society
1060 Elm St.
Tucson, AZ 85719

APPENDIX D: PERIODICALS

Aerospace Modeling, The Official Journal of the National Association of Rocketry
182 Madison Drive
Elizabeth, PA 15037.

Estes Educator News
Estes Industries
Dept. 217
Penrose, CO 81240.

Model Rocket News Magazine
Estes Industries
Dept. 217
Penrose, CO 81240.

APPENDIX E: GLOSSARY

Acceleration The rate of *change* in the speed of an object, usually measured in "g's."

Accelerometer A device for the measurement of acceleration.

Air flow The movement of air across a surface.

Airfoil A surface, such as a wing or a fin, that interacts with air flow to provide a stabilizing effect on a vehicle.

Airframe The external structure of an aircraft or rocket. This term usually includes the wings, fins, control surfaces, and the fuselage.

AltiTrak A handheld device used to determine the elevation angle of a model rocket for altitude calculation.

Amateur rocketry Rocketry activities that involve mixing fuel and fabricating rockets from metal.

Angle of attack The angle at which a wing or fin moves in relation to the relative air stream.

Apogee The point during a rocket's flight (or in the orbit of a satellite) at which it is farthest from the surface of Earth.

Armed The condition of the electrical ignition system after the safety interlock key is inserted in which the model rocket can be launched just by pressing the launch button; ready-to-launch condition of the electrical ignition system.

Aspect ratio The relationship of the wing span length to the wing chord, expressed numerically as the number of times the span can be divided by the chord.

Average impulse The total impulse of a model rocket engine divided by the duration of its thrust.

Azimuth angle A direction expressed as a horizontal angle from a given direction.

Balsa An extremely lightweight wood with a high strength-to-weight ratio.

Balsa adapter A round, tapered balsa fitting used to connect body tubes of different diameters and to provide a gradual transition between the tubes. Adapters are sometimes called "reducers" and can be made of plastic.

Barrowman Method A mathematical method for calculating the center of pressure of a model rocket.

Baseline The distance between a tracker and the launch pad in tracking with one station, and the distance between tracker #1 and tracker #2 in two-station tracking.

Bernoulli's Principle A principle in aerodynamics established by the Swiss physicist Daniel Bernoulli, stating that when the velocity of a fluid (such as air) is increased (as by flowing around an airfoiled surface), the pressure exerted by the fluid is decreased. A wing produces lift by forcing air to flow faster over its upper surface, decreasing the pressure on the upper surface while the pressure on the lower surface remains the same. Thus, the wing is "lifted" by the greater pressure on the lower surface.

Blast deflector A flat or curved sheet of metal that deflects the exhaust of a model rocket engine from the ground and the launch pad.

Body tube A specially wound and treated cardboard or lightweight plastic cylinder used to make the fuselage of a model rocket.

Booster A separate, detachable portion of a rocket containing its own engine and fins. The booster provides initial velocity to a model rocket before igniting and separating from the second stage of the rocket.

Boost glider: Sometimes called a "B/G," a model rocket, which through a change in weight distribution or configuration, becomes a glider after engine burnout. A boost glider ejects its engine after burnout, whereas a rocket glider does not.

Boost phase The portion of the flight of a model rocket during which the engine is providing thrust, "boosting" the rocket upward.

Burnout The point at which the propellant of a rocket engine has been totally consumed.

Burnout velocity The speed of a rocket at the instant the propellant in its engine has been totally consumed.

Center of gravity The point in a rocket around which its weight is evenly balanced. A rocket will balance on a thin edge at this point. Abbreviation: CG. Symbol: ◕

Center of pressure The center of all external aerodynamic forces acting on the complete rocket, including the fins. Abbreviation: CP. Symbol: ⊙

Chord The length of a straight line through the centers of curvature of the leading and trailing edges of a wing.

Cluster A group of model rocket engines working as a single unit, ignited simultaneously. The total thrust of a clustered unit, which can be composed of two, three, or four engines, is equal to the thrust of the individual engines added together.

Coast phase The portion of the flight of a model rocket immediately following propellant burnout and preceding the ignition of the ejection charge of the engine, during which the rocket coasts on its momentum. Model rocket engines usually emit a smoke trail during the coast phase, and this is sometimes called the "tracking phase."

Coefficient of drag A measure of how easily a given shape moves "through" (passes by) molecules of air.

Configuration The external shape or form of an object, or its structure.

Continuity light The bulb that is illuminated when the safety interlock key is inserted into the electrical ignition system if the electrical lead wires are carrying electrical current through the igniter.

Control surface A rudder, flap, or aileron built into a wing or fin that can be moved in such a way as to alter the surface of the wing or fin and thus interact with the air flow to alter the direction of the aircraft or rocket.

Data reduction The process of calculating the performance of a model rocket, based on observations and timing of its altitude, speed, and so on.

Deceleration The rate of *decrease* in the speed of an object, usually measured in "g's."

De Laval nozzle The nozzle used in high-performance rocket engines, consisting of a convergent section, a throat section, and a divergent section.

Delay or delay element A slow-burning, smoke-producing element that is ignited after the propellant in a single-stage or upper-stage model rocket engine burns out. This provides an aid to spotting the rocket and a time delay during which the rocket coasts to apogee.

Dethermalizer A small fuse or electronic timing device used to activate a mechanism for destroying a glider's lift after a predetermined time so that the glider will return to the ground.

Dihedral The upward or downward inclination of a wing with respect to the horizontal.

Drag Aerodynamic forces acting to slow an object in flight.

Duration The period of time during which anything lasts.

Ejection The act of forcing an object from a container, as in the ejection charge of a model rocket engine forcing a parachute from the rocket's body tube.

Ejection charge A charge in a model rocket engine that is ignited by the delay element and produces a miniature explosion which ejects a recovery device.

Elevation angle The vertical angle recorded by trackers, used to compute the altitude of a model rocket.

Engine (model rocket) A miniature solid fuel rocket motor that contains propellant, a delay element, and an ejection charge, and is composed of nonmetallic substances.

Engine block A hollow block or ring positioned in a model rocket body tube or engine mount to prevent the engine from moving forward during acceleration while allowing forward movement of the ejection gases.

Engine classification A system of coding used to designate a model rocket engine's performance parameters.

Engine mount The parts that secure the model rocket engine in place in a model rocket, usually consisting of an engine mount tube, an engine retaining hook, and two engine mount adapters to fit the engine mount tube to the model rocket body.

FAI (Federation Aeronautique Internationale) The Paris, France-based organization that is responsible for documenting all records in real and model aviation and astronautics.

Fillet A layer of glue placed at the junction of a fin or other model rocket component and the body tube or other component. The fillet adds strength (reinforcement) to the joint and, when smoothly applied and shaped, cuts drag.

Fuselage The body or main component of an aircraft or rocket.

"g" Abbreviation for gravity, which is a unit of measurement of acceleration, equal to the rate at which objects fall toward Earth, or approximately 32 feet per second per second.

Grain The direction in which the fibers in wood grow, and the pattern made by these fibers.

GSE Acronym for ground support equipment. GSE consists of the launch controller (electrical ignition system), launch pad, and associated devices or equipment.

Horizontal Parallel to the horizon; level.

Igniter An electrical device that initiates the combustion of the propellant in a model rocket engine.

Ignition The instant at which a rocket engine's propellant begins to burn.

Ignition system The electrical system including batteries or dry cells, wires, micro-clips, and associated circuits and components, used to ignite a model rocket engine.

Impulse Force producing motion.

Laminar flow Nonturbulent air movement near the surface of a body; air flow parallel to the surface, usually found at the front of a smooth body moving in relation to the air around it.

Launch The liftoff of a model rocket following the ignition of the engine.

Launch controller (See ignition system.)

Launch lug A round, hollow tube made of paper or plastic that is attached to the fuselage of a model rocket. The launch lug is slipped over the launch rod and guides the rocket in a stable path until stabilizing velocity is reached.

Launch pad The structure used to support and guide a model rocket during its first few feet of flight. The launch pad may also support the wires leading to the launch controller and includes the launch rod, the blast deflector, and a stable base.

Launch rail Also known as a "C-rail," a launch rail is a metal rail with a hollow square shape, resembling a squared "C" in cross section. This is an alternative to the launch rod in a launch system.

Launch rod A rod, ⅛ inch to ¼ inch in diameter, which supports and guides a model rocket in a vertical path until the rocket has reached a speed at which the fins become effective in stabilizing it.

Leading edge The foremost edge of a fin, facing the direction of travel of a rocket.

Lift The force produced by an object moving through a fluid that is perpendicular to the direction of motion.

Liftoff The first motion of a model rocket after ignition.

Mass ratio The ratio of the mass of a rocket, fully loaded, at liftoff, to its mass after its propellant (fuel and oxidizer) has been burned. A mass ratio of 2 to 1, for example, indicates that the rocket's propellant makes half of its weight. The lower the mass ratio, the more efficient the rocket.

Microclips Tiny, flat-jawed alligator clips used to attach the leads from a launch controller to an igniter.

Mini-engine The smallest solid propellant model rocket engine; 1.75 inches long by 0.500 inches in diameter; usually used in high-performance model rockets or for sport flying from small launch sites.

Model rocket A model made of such materials as paper, wood, and plastic, without the use of substantial metal parts, which is powered by a commercially available solid propellant model rocket engine. A model rocket rises without lift-producing aerodynamic surfaces and contains a recovery device to bring it safely back to the ground.

Momentum A property of a moving object equal to its velocity multiplied by its mass.

Multistage rocket A model rocket having two or more engines, each used during a different portion of its flight, in series. Each engine and its associated supportive structure is dropped as soon as it ignites the engine in the stage above it, which improves the rocket's mass ratio.

Mylar An extremely lightweight and strong plastic sometimes used for streamers and parachutes. Some Mylar has a silvery coating on one side, which it makes it highly visible from the ground.

NAR The National Association of Rocketry, an organization founded in 1957 to perpetuate model rocketry on an individual and organized basis, to develop standards and safety codes, to establish standardized competition, and to sponsor national championships.

Newton A force, or measurement of force, which gives a mass of one kilogram an acceleration of one meter per second per second. Abbreviation: N.

Newton-second The metric counterpart of "pound-second," which serves as a measure of impulse. A Newton-second is equal to one Newton of force applied for one second. Abbreviation: N-sec or N-s.

Nichrome A type of high-resistance electrical wire that becomes hot when an electrical current is passed through it, sometimes used in model rocket engine igniters.

Nose block (bulkhead) A round balsa fitting used to block off a payload section or body tube.

Nose cone The foremost surface of a model rocket, generally tapered in shape to streamline it, made of balsa or lightweight plastic.

Nozzle The exhaust duct of a rocket engine's combustion chamber. Gases from propellant combustion are accelerated to higher velocities in the nozzle.

Oscillation A periodic motion, such as the rolling, pitching, or yawing of an aircraft or rocket, or a combination of these.

Payload The cargo of a rocket.

Parachute An umbrella-shaped device, made of fabric or lightweight plastic and attached to an object by shroud lines, used to slow the descent of a falling body through the atmosphere.

Pitch Rotational movement about the lateral axis of an aircraft.

Pop-pod An engine pod that is jettisoned from a boost glider after the activation of the engine's ejection charge and recovered separately.

Pound-second The English measure of impulse, now replaced in general use by the metric unit "Newton-second."

Powered flight The "boost phase" of the flight of a model rocket.

Propellant The source of motive energy in a rocket engine; a mixture of a fuel and an oxidizer.

Pylon In a boost glider or rocket glider, the mount, usually streamlined, for the engine holder.

Recovery system A component of a model rocket, or a device incorporated in a model rocket, whose purpose is to return the rocket to the ground in a safe manner. All rockets *must* employ a recovery system, such as a parachute, streamer, or other drag-inducing device or system.

Recovery wadding Flame-resistant tissue that is packed between the model rocket engine and the streamer or parachute to protect the recovery device from the hot gases of the ejection charge.

Relative air stream The apparent motion of the air in relation to a body in it; motion may be caused by moving air or by a body moving in relation to still air.

Rocket glider Also known as R/G. A model rocket that, through a change in weight distribution or configuration, becomes a glider after engine burnout. Similar to a boost glider, but it retains its engine after burnout.

Root edge The edge of a wing or fin that is attached to the fuselage.

Rudder A moveable or fixed aerodynamic control surface used to cause yaw movement, or to prevent such movement, to keep an aircraft or rocket on a fixed horizontal path. B/G's and R/G's sometimes use fixed rudders to cause the model to curve to the right or left constantly, resulting in a spiral flight path during recovery.

Safety key A special key used to activate a launch system. Without a safety interlock key, no power can get to an igniter.

Scale model An extremely detailed model that duplicates in miniature the dimensional relationships and appearance of a prototype.

Shock cord The elastic cord used to attach the recovery system (parachute or streamer) to a model rocket's fuselage. The shock cord commonly reaches from inside the forward end of a rocket's fuselage to the bottom of the nose cone and is coiled up inside the body tube during flight. The elasticity of the shock cord absorbs the shock of recovery system deployment and rapid decleration.

Shroud lines The string or cord used to attach a parachute to a shock cord or nose cone.

Sine In a right-angled triangle ("right triangle"), a function of the acute angle equal to the length of the side opposite the angle to the length of the hypotenuse. Abbreviation: sin.

Single-station tracking A system of tracking model rockets using only one tracker. It is the least accurate of tracking methods. (See "two-station tracking.")

Smoke-tracking and delay element The portion of a model rocket engine that provides a timing method to ignite the ejection charge at a specific interval after the propellant has been consumed and which produces a trail of smoke to aid in observing the rocket in flight.

Specific impulse A measurement of propellant efficiency, expressed in seconds, and determined by dividing the total impulse of the engine by the weight of the propellant. The higher the specific impulse of an engine, the more efficient it is. Most model rocket engines have a specific impulse of between 50 and 100 seconds.

Stability The tendency for a rocket having a proper CG/CP relationship to maintain a straight course along the axis of its thrust (opposite the direction of its thrust) despite rotating forces caused by outside disturbances.

Stabilizer A fixed, or unmoving, stabilizing surface, usually mounted on the tail of an aircraft in a horizontal position.

Staging Igniting engines in sequence, each dropping off after its propellant is consumed and the engine above it is ignited.

Standard engine A model rocket engine of the usual size; 2.75 inches by 0.690 inches in diameter.

Streamer A recovery system using a long, narrow piece of flame-resistant crepe paper or plastic (such as Mylar) attached to the nose cone or shock cord of a rocket.

Sustainer engine A single- or upper-stage engine equipped with a delay element and an ejection charge.

Tangent In a right triangle, a function of an acute angle equal to the ratio of the side opposite the angle to the side adjacent to the angle.

Theodolite An optical instrument for precision measurement of horizontal and/or vertical angles.

Throat The portion of a rocket engine nozzle having the smallest cross-sectional area.

Thrust phase The period of time during which the propellant is burning and the rocket motor is producing thrust.

Turbulent flow Uneven air movement over the surface of a body in which the air movement is not smooth but has ripples and eddies, usually around an uneven surface.

Two-station tracking A tracking system employing two trackers.

Velocity A rate of motion in a given direction, measured in terms of distance moved per unit time.

Vertical Perpendicular to the horizon; up and down.

Weathercock To turn into the wind, away from a vertical path.

Wind tunnel An enclosed chamber through which air is forced at varying velocities, used to test the stability and flight characteristics of aircraft, rockets, and aerodynamic components, in model or full-size form.

Wing The main lifting surface of an aircraft.

Yaw Rotation of an aircraft or rocket about its vertical axis.

INDEX